R. E. Francillon

National characteristics, and flora and fauna of London

R. E. Francillon

National characteristics, and flora and fauna of London

ISBN/EAN: 9783743323520

Manufactured in Europe, USA, Canada, Australia, Japa

Cover: Foto ©berggeist007 / pixelio.de

Manufactured and distributed by brebook publishing software
(www.brebook.com)

R. E. Francillon

National characteristics, and flora and fauna of London

NATIONAL CHARACTERISTICS,

AND

FLORA AND FAUNA OF LONDON.

BY

R. E. FRANCILLON,

AUTHOR OF "PEARL AND EMERALD."

LONDON:
SMITH, ELDER AND CO., 15, WATERLOO PLACE.
1872.

PREFACE.

THE following short sketches bear sufficient signs of their ephemeral production. This I put forward as an explanation rather than as an apology. The "National Characteristics" are meant, not to be exhaustive, but to bring out some salient points in ourselves and in our neighbours to which habit and conventional judgments are apt to blind us. It is not my fault if they sometimes appear paradoxical. The "Flora and Fauna of London" is a slight attempt to grasp some portion of the poetry that clings far more to city bye-ways than to country lanes. London is the greatest of all poems: and it is well to suggest how the pilgrim of the flag-stones may find food for fancy without spending so much as a penny at a book-stall.

CONTENTS.

—◆—

NATIONAL CHARACTERISTICS.

FLORA AND FAUNA OF LONDON.

NATIONAL CHARACTERISTICS.

NATIONAL CHARACTERISTICS.

I.

ALPHONSE.

" COULD some power the giftie gie us, to see ourselves as others see us," we should see— something altogether unlike our real selves. A *Meess Anglaise*, with her flat ringlets, her long waxen face, and projecting teeth, is—in spite of Gavarni, of the *Charivari*, nay, of M. Taine himself—not more unlike the English girl than is the Frenchman whom tourists see through their travelling spectacles unlike the Frenchman whose real image is reflected upon the naked retina, but not conveyed to the brain. We are, perhaps, more generous in our judgment of our neighbours

than they are of us ; but, unfortunately, even
envy does not mislead so much as generosity.
The classic saying, " They do these things better
in France," which Sterne, the original ancestor
of English humour, meant to be taken—not
exactly literally, is a leading case in the common
art of misunderstanding irony.

Hogarth has much to answer for, besides the
mistake of thinking that the idle apprentice always
comes to the gallows, and that the industrious
always becomes Lord Mayor. The anatomical
caricature whom he, and no officer of the guard,
put on duty at Calais gate, is not the least like
any French soldier that ever was or that ever
will be. The scarecrow refugees whom he depicted
tripping out of church in their land of refuge
were not, one may be sure, the sons and daughters
of the Huguenot gentlemen who fought so fiercely
and so stoutly for their faith in the Cevennes.
But Hogarth was an Englishman : and so to
this day, even those who have with their own
eyes beheld the real French soldier, strong, active,
well-fed, and broad-shouldered, though not rivalling
the lofty moving tower we call a Guardsman, will

still talk with contempt of his physique as though Waterloo had been an easy victory to win, or the Malakof a playground. Englishmen seem to need English air for the proper use of their eyes. There are very good people who have lived for years at Tours or Boulogne, and who have not been able, in spite of all their experience, to free themselves from the strange delusion that our nearest neighbours are distinguished for wit and polish above all nations of the civilized world. A lady has been known to travel from Calais to Marseilles, a butt for stares such as only your true-bred Frenchmen can stare, and a victim of frantic crushes at *tables d'hôte* and elsewhere such as only a herd of hungry Frenchmen can form, and yet hold to the nursery legend that these monsieurs are the very pink of politeness, and her own countrymen a race of Polar bears. It is very strange. After much consideration, a long course of reading, and careful comparison, I have come to the conclusion that the conventional Frenchman is something as follows :—His name is Alphonse, unless it happens to be Jules. He is slight, meagre, but graceful in make, olive complexioned, with white teeth, and a good crop of glossy hair, worn *à la* porcupine. He is quick in speech, and gesticulatory in manner; frivolous, perhaps, but lively, witty, gay, and ready at repartee; careless of time and money, and not over moral in his domestic relations; but courteous to all men, and gallantly, even chivalrously, polite to *les dames.* His daily life he has made so agreeable, down to every detail, that he may be excused for not being able to translate the word Home. His taste in every art, up to that of cookery, is delicate and refined, while his quick sense of honour makes him as considerate of others as he wishes others to be of him. Laugh at him, caricature him as we may, these traits still lie at the bottom of our loudest laughter, and our grossest caricatures. *Le voilà,* the Frenchman. *Le voilà le Chameau!* It is all as false as perjury itself, every word. Behold the true Alphonse, as he is—which is not quite the same as what he ought to be.

Alphonse—why do we always think of him as Alphonse?—is so far from being slight or meagre that he achieves the dignity of *embonpoint* when broadshouldered young Englishmen can still wear a lady's girdle. I am talking of the typical Alphonse, be it understood, not of the few million exceptions that are, of course, to be found here and there. So far is he from being specially favoured in point of *chevelure*—though I admit the porcupine—that a gathering of Frenchmen of any class, to a spectator enjoying a bird's-eye view, in no way differs from one of British *savants*, who have a traditional right to the smoothly-polished crown. There is no reason to think that Parisian dentists complain of slackness of trade, while the olive complexion—which ought, by the way, according to literal interpretation, to be a shade of dull green—is about as common north of Provence or Gascony as in Middlesex. In figure, he is less inclined to be graceful than clumsy. Under these circumstances, vivacity can scarcely be expected ; and Alphonse—not the conventional Alphonse—is not vivacious. His soul is too much engrossed with the funds, with the shop, with the grim politics of the transitory minute, to have much room for *esprit*—the delicate bouquet which vanished for ever in the foul reek of the guillotine eighty years ago. A witty Englishman is a rare bird, but a witty Frenchman, whatever may have been the case in by-gone days, is a veritable black swan. As to humour—tell a good story, such as ears may hear without reddening, to our typical Alphonse, and you will see. It is France, not England, which is the true *nation bourgeoise*, the nation of shopkeepers. Alphonse, our typical Alphonse, is eight times out of ten a shopkeeper in some form; nine times out of ten he has the soul of one. He is economical to niggardliness, and takes care not only of the francs, but of the centimes, and pockets his lumps of sugar. The *flâneur*, the *petit crévé*, is not the typical Alphonse—he works laboriously for his dull pleasures, and measures his satisfaction in them by the sous he saves. And, alas! it is not Alphonse who makes room for you at table, or who apologises for having trodden on your train— it is your compatriot, whom you have nevertheless labelled "Bear." And gaiety—is it found among the children who do not know how to play, the young men who know how to play only too well, the young girls who are buried out of sight until the day of sale, or the keen-eyed business women, who make such admirable commercial partners in maturer years ?

It would take a volume to trace Alphonse in all his Protean forms—to analyse him, and to point out where, how and why the rule ceases and the exception begins. He, too, has his varieties—his types and sub-types, his orders, genera, and species. Only, when you meet him, see which portrait he most resembles—that which I have drawn, or that of his great grandfather, who died long ago, but which we still obstinately regard as of more than photographic fidelity. Why will we insist upon keeping up a fiction of the *ancien régime*, with all graceful vices and buried graces ? Perhaps in time we may get rid of the delusion that, while our labourers and artisans drink ale and gin, Alphonse, *ouvrier*, is content with *eau sucrée*, or with harmless and wholesome wine. Perhaps we shall one day learn that national resemblances, at any given period, are more numerous than national differences, and that individual peculiarities are no index to the characteristics of a people.

It is hard to say of a nation that it has lost its race of gentlemen. But of France it must be said that her great and still progressing revolution has sacrificed her *caste* of gentlemen—such a caste as once, with all its vices, made French society synonymous with grace, brilliancy, and courtesy. Assertive self-consciousness has taken the place of well-bred ease, and the conventional traditions are built upon a foundation of shifting sand which may be, or may not be, transforming itself gradually into a rock of rough and self-reliant virtues. It may be hard to give up our picturesque Alphonse of tradition, and to take another in his room. But the true Alphonse is a fact, and a very solid fact besides ; and it is not a little ridiculous on our part to insist upon setting up the ghost of his grandfather in his place, and calling it the Frenchman of the present day.

II.

HANS.

IF there was one person in the world whom we thought we knew, that man was Hans. How, in fact, should we fail to comprehend our own first cousin? Not that we were very proud of the relationship. We had laughed at him and sneered at him ever since the Reverend Peter Pindar, who did not reverence many things, took to flying at such small game as Madame Schwellenberg. Of course we knew him. It was nothing to us that there were such distinctions as Swabia, Austria, Saxony, Brandenburg, Hans was Hans, whether in lively Vienna or un-lively Berlin. There have been benighted foreigners ignorant enough to imagine that an English milord habitually appears in the park with the kilt and philabeg of old

Gaul. Is it wonderful, therefore, that we who of course are neither benighted nor foreigners, should, model all the *têtes carrées*, as the French call the Germans, upon one type, and that our typical Hans should be as conventional as the stage Irishman?

Let us endeavour to clothe with flesh that ghostly Hans. The task ought not to be hard, for it would be hard to find his bones. He is something of the build of the conventional Low Dutchman—who, by the way, thanks to Alva's soldiers, is often as much Spanish in feature and complexion as a peasant of Connemara. Hans is of low stature, pale and flabby complexion, colourless eyes, protected by spectacles, and scanty flaxen hair. He smokes Knaster by the hundredweight, and washes it down with butts of Bairisch beer. Parallel with these occupations he dreams over transcendental philosophy, which, to our common sense, seems like dreaming over the smoke that curls slowly from his china bowl. He was a hot duellist in his short span of youth, and wears the seam of a sabre cut that he will carry to the grave. He is a pattern husband, and his great holiday is to give to little Hans and Gretchen the treat of a Christmas tree. Of course we do not forget that Germany is the land of learning and of intellectual culture. But we distrust the learning that ends in a Strauss, and the music that develops into a Wagner. We are speaking of the type that rises before our eyes when we hear the name of Hans, the product of *brat-wurst* and *sauer-kraut*, of Kant and Jean Paul. There was once a celebrated professor who had never put on his boots for twenty years. We feel certain that professor's name was Hans. Some of us despised him, but without hatred ; others liked him, and his quiet, kindly, homely ways, but made him a sort of national butt all the same.

Suddenly—could it be that we were wrong ?—there marched across the Rhine a man, wearing spectacles it is true, and of cold, not to say repellent manners. But he was as tall, as broad-shouldered, as firm on his legs and as strong in his arms as a picked English grenadier. His characteristic was a crushing force of will. " Without haste, but without rest," as Göthe says, he had

made up his mind that something had to be done,
and he did it, with no pause for dreaming by the
way. He was an intelligent machine, rather than
a man ; superior to wear and tear, to hunger and
fatigue. He had learned not only strength, but
its economy. Nothing unnecessary he did : but
he stayed at nothing. On principle he could be
deliberately cruel, and did not even soften the
weight of a heavy blow by one apparent touch of
human sympathy. We knew, for a fact, that this
was the real typical Hans—the noble, the peasant,
the artist, the shopman, the student, the professor
—all were there, and yet, to look for our old Hans
was to look in vain. He was prompt, practical,
hard, energetic, sudden, thorough. Could it be
the same? Or was this a startling instance
indeed of the mutual ignorance between nations
of the same blood, and almost of the same tongue?

It was, and it was not the same. The Hans
whom we had known had never been—he had
been the exception, not the type even of so much
as a university town. We must remember that
nations are, after all, but aggregates of men. On
that principle, if Hans has been a dreamer, his

dreams have been very like other men's waking.
Germany, that we have deemed so slow, is the
swiftest country in the world, therefore is Hans
the swiftest of men. Go to the Latin races to
look for dreams and dreamers. German literature,
for instance, equals, if it does not more than equal,
our own. Yet ours has been the slow growth of
more than three centuries—the German, the rapid
growth of but little more than one. German
political life has almost reached maturity in half
the time. Heine, who was not a German, and
who had the contempt of a foreigner for Hans,
taunted him with his slowness in not doing and
becoming all things in a day. Little cared Hans
for the sharp lashes of the witty Jew. He was
growing in the stem—not shooting out random
branches that are cut off before fruit can come.
The typical Hans has, in short, a perfect mania
for the practical. All he does, all he thinks, all
he reads, all he dreams—if he must be said to
dream—are tasks to which he sets himself with a
deliberate end in view. It may be that his object
is to save a competence, it may be that it is to
trace the derivation of a word, or to enjoy himself,

or to swindle, or to invent a system of theology.
But in any case he is not content till he has gone
as far as the day's length will permit, and the
day's task is done. He has none of the cynic's
humour. All things are worth doing, and all
things must be done thoroughly, or not at all.
No wonder the typical Hans has few words to
spare. No wonder that he is self-absorbed—not
to say selfish—in his paths and in his aims. Add
to this that he has, as a rule, the physical endu-
rance of an elephant, and you have described
Genius — something beyond our mere common
sense, however we may boast, to understand. No
wonder that Hans runs into cloudy sentiment as
a relief from hard realities, and seeks rest from
labour in fanciful speculation. It was even so that
the First Napoleon chose for his mental recreation
the poems of Ossian — at first sight the most
uncongenial to such a mind. No wonder, too,
that, when brought into contact with lighter-
hearted and brighter races, the hand of Hans
feels like a cold grasp of steel. It crumpled up
the petals of Italy as Götz of the iron hand
would have plucked a flower. As for his home-

liness what home is that where the wife is but cook and stocking-mender, without an idea in common with her lord and master save that of work, which, in her case, degrades her out of the bright intelligence and gracious ways that ought to be the birthright of every woman in the world ? The home of Hans is in himself : he does but eat and sleep in the warmth of the stove. There is no man so falsely represented, even in his own literature, as Hans. His literature is so purely poetic just because he himself is such pure prose. He cannot write in prose ; but he can live it, and he lives it sternly, with no thought of comfort as an end, or of turning aside to gather the few flowers that grow by the way. If they come of themselves, it may be well ; if not, perhaps all the better.

On the whole, I consider that Hans is a man who first of all loves an idea of some sort, then himself, and then all things and persons that are serviceable to either of the two. That he is cold, not by nature, but because he is self-absorbed. That he is inclined to sentiment and to poetry because he is practical to the very core. That he, as long as he can live his own life, is careless of what others think of him. That he is great in art because art also is a form of work, and therefore worth the doing just as much as digging in the fields. Lastly, that all his other qualities, good, bad, and indifferent, are developed from these. Of course I am keeping to the typical Hans: nor do I refuse to regard the millions of his brethren who are different from him in the details that everywhere distinguish man from man. But at the apparent, not real, transformation of Hans we can wonder no more. Men and nations alike are capable of all things when actuated by an all-mastering and living idea, and when they have the will and the power to do with all their might all that their right hand finds to do, for the sake of doing it, and not for the sake of daily comfort or empty fame.

III.

GIUSEPPE.

THE average Englishman of Goldsmith's days may be excused for judging the Italian from the specimens who bore upon their inflated shoulders the whole weight of the lyric stage. No doubt Lord Allcash, in *Fra Diavolo*, is a very exact portrait of an Englishman. It is a well-known habit of Englishmen to travel in curl papers and green veils, with all their wives' diamonds packed up in a hand valise for security, and to make "Oh, yes," and "Shocking," do duty for conversation. It is also well known that an Italian is a gifted, but wayward being, who lives a life of melody and maccaroni; who is either a Papal fanatic or political conspirator, according to circumstances, and who translates *summum bonum* by

2

dolce far niente. It is supposed to be rather a
privilege to be acquainted with a "signor"—why,
it is hard to say, unless a "signor" is a remarkable
and interesting creature by right of not being
either Mr. or Monsieur. At the same time it is
the fashion to conceive that no Englishman can
understand an Italian: that while we represent
hard, prosaic common sense he represents the last
feeble flash of whatever poetry still lingers as a
reality among men. And yet we cannot forget
either historic Italy, or be blind to the active life
that seethes in her to-day. No wonder that the
habitué of the Opera is confused. In effect, we
have made up our minds to decline Italia according to the rules of the feminine declension. She
has been called the "Niobe of Nations;" and a
weeping statue she shall remain, in spite of everything. Giuseppe—we are not so far out in our
typical name — shall be a maccaroni-munching
lazzarone, a dreamer, subject to the malady of
eternal childhood. No wonder that we do not
comprehend him and his ways, for we will not see
the glaring fact that there is no one so like an
Englishman as the typical Italian—be he from

Rome, Naples, Florence, or Turin. No one is so like the typical Italian as the typical Englishman.

We ought to be ashamed of being imposed upon by tricks of manner, which are no more part of the essence of the Italian than our tricks of manner are of ourselves. The Neapolitan's warmth of nature is no more to be measured by his warmth of manner than rigidity of tongue or limb is to be taken as the index to the true nature of the Englishman or Piedmontese. Had we all of us Giuseppe's flashing dark eyes and mobile mouth —mere accidents that they are—we also should show all our passing impulses in the same demonstrative way. It is mere slander to call the Italian insincere. Only he unconsciously shows his transient feelings, while our grey eyes are incapable of showing those that take their root in our souls. It is absurd to say that we Englishmen have not our full share of insincerity—it is nothing more than that climate has given us the advantage of being unable to express as much as we wish to be thought to mean. There never was a dark-eyed man yet, of whatever nation he might be, who was

not open to the charge of insincerity; never a
grey-eyed man who was not held to be cold and
reserved : while, in reality, the true difference
between them might lie all the other way. And,
as with men, so with nations, which are made up
of men.

It is true that the Italian forgets his protestations of eternal devotion when your back is turned,
and he finds himself in a position to make the
same protestations to your deadliest enemy. What
would you? It is nothing more than oiling the
hard wheels of human life to be on comfortable
terms with all. It is true that with the traditional
esprit fort of his country—has not Italy been the
cradle of free thought since the days of Dante ?—
he retains the superstitions of the nursery. Is it
not well to stand well with all the powers that are?
He keeps a conscience, and finds it sometimes
clash with nineteenth-century doctrines of liberty,
to which his living eyes cannot be blind, so he
temporises, and if he can no longer believe in the
Pope he follows Garibaldi, and sends for the
priest when the physician of the body warns
him that he must die. But in all these things

does he differ so widely, in principle, from ourselves.

There is but little of the suppleness with which he is credited in the nature of the Italian. It is on principle that he seeks to be all things to all men. Know him well and you will find that he has a strong individual character of his own. He is not like the natives of *la Grande Nation*, of whom, if you know one, you know all. He is in the first place an Italian, then a Milanese, Venetian, or Neapolitan, then the offspring of a parish, different from all other parishes; and, lastly, the Giuseppe who differs from all other Giuseppes as much as the letter A differs from the letter B. His sole national characteristic is that, unless he comes from the north, he has dark hair and bright eyes. In all other respects, you must know Italy as you know England before you can pretend to say that you know in what way he differs from you. You may know an Italian, but the typical Italian you cannot know. The typical Italian exists nowhere beyond the pages of *Childe Harold* or the walls of Covent Garden. He is not Fra Diavolo, and yet he may be Fra Diavolo, and a great deal more.

He is not, typically, the Garibaldino, nor the Cordino, nor the tenore, nor the lazzarone, nor— but what end is there to the catalogue of what he may or may not be ? The Italian may be anything. He may be the professional *cavalier serviente*, who spends his days in saving another man's wife the trouble of flirting her own fan ; or he may be the reckless soldier, or the devoted priest, or the *chevalier d'industrie*, or the operatic tenor, the brilliant poet, the flashy poetaster, the astronomer, the political economist, the engineer of Mont Cenis, the Mazzini, the Cavour. The typical Italian is as versatile as the typical Englishman. How therefore can we understand him any more than we can understand ourselves ? Giuseppe is as tall, as broad-shouldered, as strongly made as an Englishman—" barring the beef." He is cool and steady in fight, without the boasted *élan* of the *grande armée*, but without the tendency to panic which is the natural result of *élan*. He is our rival in respect of the grosser forms of *gourmandise*, and if he stays his appetite with rice and maccaroni—why, if we had not good mutton, we should become maccaroni eaters too. It is a mistake, moreover, to accuse Giuseppe of being an imaginative creature. There are but two countries in the world wherein common sense reigns supreme over logic, reason, experience, all sense, in short, which is uncommon—and these are England and Italy. An Italian—the typical Italian, I mean— thinks more of taxes than of ideas, and, in spite of conventional tradition, amuses himself very much on system, as though, as with us, amusement were a natural part of the day's toil that has to be got through. He is theoretically fond of change, but slow to improve, and his first question always is, Will it pay or save? Giuseppe is a born economist, at least in the matter of half-pence : and if he only possessed a little more national energy—but the parallel is interminable. I cannot help feeling that an Englishman should be shy, not to say humble, in judging Italy. It may be patriotic to modestly confess our inability to understand, but it is, at the same time, to deny a bond of sym- pathy which ought to be the pride of the two greatest nations of the modern world. Every sarcasm launched at Italy is reflected from the cliffs of Albion. Florence is a home to Englishmen—London to Giuseppe, such as to neither can be Paris or Berlin. Nor have we, with all our wealth of iron and coal, carried out one idea— steam apart—which had not its birth in the land which was once the appanage of Rome.

IV.

JUAN.

CERVANTES, Le Sage, and Beaumarchais are our
authorities for the manners and customs of Juan. It
is observable that very few people, comparatively,
are acquainted with a live Spaniard, so that the
fellow-countrymen of Cid Rodrigo y Diaz share
with their national hero a quasi-poetic character. Once upon a time not to know Juan
was much the same as not to know anything
beyond the limits of a country town. Spain was
to the far North, South, East, and West what
England is now; and even more, for it was rather
easier to hold the world in one's hand than it has
become since the days when the map of America
was more barren than that of Central Africa.
But we do not take for our typical Juan the ship-
mates of Columbus and Pizarro, the soldiers of

Alva, or those who fought either in the Bay of Lepanto or—with not quite such good success—in our narrow seas. Of course it is necessary to prefix to his name the title of "Don," and to translate his name into the Italian Giovanni. The ladies of Cadiz, whom Childe Harold found so fascinating, would, it is needless to say, scarcely fail to bring flirtation into fashion if there was ever a time or country in which it was a thing unknown. Now it is absolutely necessary that the lover, whether of romance or comedy—and it is there alone that we find our typical Juan—should be furnished with a guitar, or at least with a mandoline. So a guitar we accordingly give him, and even idealise that very sorry instrument until its feeble tinkle becomes poetical, if not very musical. With that instrument we send him forth to conquer, not the Indies, but Donna Inez, who, as all the world knows, likes nothing so well as to be woke up at the cold hours of morning to listen to a serenade. In personal appearance Juan is decidedly interesting. He is glued to a cigarette, and lives upon garlic and pumpkins. He earns his living by shouting "Bravo toro!" at bullfights: so that, according to the laws of trade as they exist in more prosaic lands, his tailor must be content to wait a considerable time before he is paid for that gorgeous costume *à la Matador.* The intrigues in which he is hourly engaged would drive an ambassador to despair; out of gestures and fan-flutterings he has constructed a code of diplomacy far superior to that of the Atlantic Cable itself. At the same time he is grave, dignified, and as ready at the knife as the conventional alderman. In fact, if all accounts be true, the land of Juan is a *rechauffé* of a great many things, which, except the garlic, are very charming and very impossible.

A visit to Seville is, it must be owned, a little disappointing: a visit to Madrid still more. Even as the Englishman who has carefully trained himself out of books in the accent of the most "choice Castilian" finds it anything but the pronunciation recognised in Castile, so does the reader of tradition find himself, though his voyage is over, still at sea. *"Cosas d'España"* —things of Spain—are peculiar; but they are

not much more peculiar than other things elsewhere. Juan is there in the flesh, but the spi-
ritual charm is gone. One very soon finds out
that what is most beautiful at a distance may
become not only very unpicturesque, but very
disagreeable. A beggar boy by Murillo is a
magnificent study: so is a monk by Velasquez;
and at a fancy ball, on an English brunette, there
is nothing so charming as a Spanish costume.
But Murillo and Velasquez were painters great
enough to have found equally good subjects to
idealise Whitechapel, and Donna Inez thinks
nothing so becoming as a bonnet fresh from the
boulevards. As to Juan himself, it is true that
he is fond of tobacco and content with the worst
cookery in the world—when he can get nothing
better, but I have never found that his forte is
sentiment, or that he is able to get his living
out of bull-fights and serenades. He may be
always in love, but he for the most part shows
it by lounging over a newspaper at the *café.* From
the poetic atmosphere which hides his country
from the outer world he inhales not a single
breath. To him the history of his country dates

from the to-morrow which never comes; and, on
that ground, he considers himself nothing if not
a politician. He belongs to whichever of the
thirteen parties his need of dollars leads him to
choose, and spends the dollar, when he gets it,
not in feeing Donna Clara's waiting maid, but in
making experiments as to how far a dollar will go.
He is—it must be said—a rather cold-blooded
animal. His eyes still flash, but the good rich
Arab blood has long begun to turn thin—or, as he
would prefer to call it, blue. There is something
terribly unsatisfactory about Juan. Everything
about him seems to turn mean and sour. By
politics he seems to understand the art of better-
ing himself at the expense of his neighbours, and
it is doubtful if he is really moved to enthusiasm even by the fullest rhetoric in which poli-
ticians of his stamp are the most prone to indulge.
He lives temperately and poorly, not because he
is a philosopher, but because all his desires could
be wrapped up in a packet no larger than a
cigarette paper—and that there is no symptom
of largeness of soul. Being human, he wants
something: but he does not know what he wants,

so he welcomes anything like a change—even a game at revolution—in order to escape from the national *ennui*. I am aware that I am flying in the face of all tradition when I call Juan the type of the lymphatic temperament pushed to its utmost verge. It is not his soul that speaks in his warm skin, his dignified aspect, and his flashing eyes. These come down to him from other days when his fathers the Goths were struggling for supremacy with his mothers the Moors. Juan is the degenerate descendant of a northern race left to decay in an uncongenial land. But tradition, bestowed by genius, clings : and no doubt every one who has visited Juan at home has looked for what he expected to see, and found it accordingly. It is very easy to create poetry everywhere, far more easy than to read plain prose—and Juan, if not Inez, is very plain indeed. Let us by all means enjoy the dream that the fairy land of Don Quixote is the Spain of to-day, and therein build a castle before which Juan may pose gracefully with his guitar. The real Castle of Indolence, wherein Juan really dwells, has been built by far less artistic hands.

V.

A L I.

THE genii whom children raise are not those such as the child James Watt raised, like the fisherman of the *Arabian Nights*, from the spout of an innocent tea-kettle. They are those which swarmed round the Sultan's couch at the magic words of Scheherazade. They swarm round the couch of Ali still, whether he dreams amid the roses of Damascus, the bales of Bassorah, the minarets of Constantinople, or the fountains and love-songs of Ispahan. Five concentric circles enclose the chamber of Ali, and hide him from our waking eyes. First come the houris and peris, with finger-tips rosy as those of Aurora, with faces like the full moon. The second is a circle of comedy, of calendars, of hunchbacks, of barbers through whom walks the great Caliph Haroun, attended by Giaffar and Mesrour. The third is composed of all who have sung and written in the East when the East was young: the fourth, of all who have sung and written of the East since the East has grown old. The fifth is the circle of the Djinn, who have made the Orient an oasis in a desert of steam, and who make a man's first sight of the Golden Horn an era in the history of his days. And it is through these misty mysteries that we look upon Ali; he is the magic centre, the type of what the child, holding Haroun's hand, and voyaging with Sinbad, will not allow the man to disbelieve in his maturer years.

Withdraw, then, shapes of steam made black and foul with the price and prose of coal, and let Ali emerge from his chamber into the outer day. What guise will he take but one, whether he be Turk, Arab, or Persian, emir, barber, or slave? His head will be swathed in the folds of the sacred turban, his figure will be draped in robes, his beard will hide his breast, and the jewelled hilt of his curved scimitar will be ready to his hand.

Bismillah ! Sit down by him on the divan, whence he never rises, but reclines all day long in dignified repose : let him clap his hands that the attendant negro may bring coffee and the chibouque : and let us find in the flesh the type of that wondrous East which, as it has been, so is, and ever will be. Thrust aside that stiff figure, in frock coat and fez, who stands in the way as you advance with shoeless feet over the floor, and who looks as though he had taken an ale-cork for the glass of his fashion and the mould of his form. He is smoking a cigar : you came to inhale the dreamy fumes of the narghilé. He addresses you in French—he must be the dragoman. You ask for Ali. Alas ! the magic rings, iris-hued as soap bubbles, have vanished into air. The ale-cork is Ali.

He has not forgotten his old-fashioned Eastern courtesy—the best, if not the most sincere, form of courtesy in the world. He will not just yet introduce you to the ladies of his family, but he will ask you to dinner, and go with you to the opera afterwards, where you will hear the " Trovatore " performed by singers who have not

succeeded too well in Naples or Milan. He will
talk to you of progress and the Eastern question
as intelligently, and almost as intelligibly, as an
habitué of Pall Mall. And yet there is an inde-
finable something about him, as about his here-
ditary enemy the Russian, that makes you feel
that if he is not the Ali of whom you dreamed, he
has only put off the turban without donning
such a hat as your own. The fez is a com-
promise. His limbs are stiff and awkward in
the very shooting jacket in which we feel most at
ease. There is a buttoned-up appearance about
him, and a straightness, as though he were
momentarily afraid of giving way. I am speaking,
of course, more particularly of the Turkish or
dominant Ali, for though the tendency is true of
all, there is no railway yet to Teheran. The fact
is, the Turk feels himself standing on a thin plank,
on which he must pose, as it were, on one leg, in
order to retain his balance, and practise the goose-
step in order to keep up a semblance of getting
on. Ali is trying to grow young, and thinks that
by a severe process of tight lacing with red tape
he will reduce himself into Occidental formality.

Only it will not do. A man does not imbibe the
Western spirit by suddenly finding out that
champagne is not forbidden by the Koran. The
Indians of the West were not put in sympathy
with their persecutors by force of fire-water, nor
did that pioneer of civilization enable them to hold
what was once their own. There is no force but in
race and in being true to its traditions, not only in
spirit, but in form. It was the old Ali, almost
such as he was when Mahomet launched him from
the desert, that needed a Sobieski to drive him
from Vienna, and a Charlemagne to keep him
beyond the Pyrenees. The new Ali is something
of an *esprit fort.* He can read, and he reads the
newspapers. He does not despise the Giaour—he
hates and tries to ape him. He has become useful
as a kind of feather-bed between the Levant and
the Black Sea. Even his wife no longer wonders
at the strange and distorted figures of her Western
sisters : she, too, has heard of a *pannier* and a
chignon, if all tales of the harem be true. If she
retains her veil, some sceptics are inclined to think
it is on the same ground as that on which certain
ladies were once supposed to go masked to the

playhouse, to conceal something else than beauty.
She will soon learn the compensatory qualities of
a bonnet of the newest fashion, and then the veil
itself will go. Jealousy is not a modern failing,
and it is not so easy to drop a sack silently in
the Bosphorus as it used to be. Gas has thrown a
cruel glare over the mysteries of the Golden Horn.
But, as I have said, race is race. To cease to be
one thing is not to become another. If London
were to ape Constantinople the effect would not be
more grotesque than for Constantinople to mimic
London. The East and the West cannot sympa-
thize, and for the East to strive with success after
Western development is as hopeless as for the
West to endeavour to stand still. It is only to
substitute frivolity for earnestness, the fez for the
turban. It depends upon the feelings of each of
us to decide whether we would for the sake of
romance retain the East in all its hideous glory, or
whether, for the sake of the triumph of steam, we
would let it sink into the realms of historic
mythology. For my part, when I wish to journey
Eastward, I take M. Galland for my dragoman,
and shall not be altogether content when a railway

junction is opened at Sinai. A pilgrimage should be something real, not a thing to be managed by a tour-monger as if it were a cheap trip to Herne Bay. However, like all things else, Ali must be taken as he is found. Only it is not merely a matter for sentimental regret that what he is should be no longer what he has been. He was the natural guardian of all that is most sacred in the history of the world. He has become an undeveloped stoker. The steam genii are closing round him, and before long, we may expect to see the lands he should for the world's sake guard from Western desecration turned into another paradise of railways and hotels—as real as our own, and perhaps managed almost as badly. When that time comes, I do not know that the downfall of the Crescent will have been an undiluted advantage. It at least admitted the fact that there are a few matters in which the world is at its best when standing still.

VI.

JONATHAN.

THE typical American is called Jonathan because
it scarcely ever happens that Jonathan is his real
name. On the same principle, called by ancient
grammarians and modern pedants *lucus a non
lucendo*, he is also popularly termed a Yankee,
because a certain proportion of his fellow-citizens
happen to be born in the half-dozen little States
of New England—the only territory whose inhabi-
tants have a recognised right to that illustrious
title. Let him hail from Tennessee, Louisiana,
even from California, he is still a Yankee. Even
so a native of old England, whether from Cornwall
or Northumberland, is of course a Londoner, all
the same. With a becoming scorn of transatlantic
things, we ignore distinctions whether of race or

of geography, and take the great Elijah Pogram—
an institution of course totally unknown among
ourselves—as the typical specimen of those whom
Englishmen, when in an affectionate mood, call
brothers, on the ground, presumably, that they are
mainly of Irish or German extraction. In short,
the typical Jonathan, dressed like a scarecrow in
the rags of a humour that is ever fresh though
ever stale, is about as true to nature as John Bull.

There is no need to repeat the work of carica-
turists by describing the typical " Yankee," who
stands alike for the New York Knickerbocker, for
the peaceful graduate of Harvard, for the bowie-
bearing backwoodsman, for the representative of a
" First Family of Virginia "—for all the types and
classes that go to make up a nation as varied as
the whole of Europe put together. We know him
by heart, and his conventional eccentricities form a
literature and a picture gallery by themselves.
The strange part of the matter is that while most
of us have been acquainted with compatriots of
that bird whose home is in the setting sun, few of
us have found in him the offensive conventionality
made up of one part maniac, one part charlatan,

one part boor, and one part rogue. Beyond
certain peculiarities of speech, dress, and accent,
which are peculiarities because they are not our
own, and certain political notions which are wrong,
because England—as some think, happily—is not
yet an outlying territory of the Great Republic, we
have been in each case content to admit that our
acquaintance afforded a piquant and pleasant
exception to the manners and customs of the
Jonathan in whom, nevertheless, we are deter-
mined to believe. Be it so. The travelled
American whom most of us have met may not
be a fair representative of the American at home.
What are the characteristics which, without regarding subordinate types, distinguish the true
American, be he Yankee or not Yankee, from the
rest of the world ?

The answer lies in one word. Barnum was an
American, and the Americans believed in Barnum.
Joseph Smith, who invented Mormonism, was a
citizen of America, and the Americans—at least
a great many of them—believed in Joseph Smith.
Spiritualism and all other words of a similar termi-
nation, free-love, and all other institutions with a

similar prefix, talk valued in proportion to tallness, promises believed in proportion to impossibility of performance, hero worship of men who are not heroes, theories without facts and imaginary facts not to be explained by theories, are essentially of transatlantic growth and transatlantic cultivation. The Land of Jonathan plumes itself upon its sharpness. It should rather pride itself upon its childlike innocence—upon its verdant credulity. Charlatans do not flourish where dupes do not abound. There is nothing that Jonathan will not greedily swallow, from a new system of cosmogony to a manufactured mermaid, a woolly horse, or a belief that he is far too smart or 'cute to be taken in. The last is the well-known craze that brands personal, and therefore national credulity. We do not plume ourselves upon our smartness. We, in our modesty, thought it was ourselves who were to be the dupes of the Indirect Claims — not the voters who did not see through that most transparent of election " dodges." But it is not we who try to scrape wooden nutmegs. We call believers in Fanchette, and in the wretched *revenants* who talk in chairs and tables, by their proper and not complimentary names. It is we, alas! who send a supply of Scotch journalists, Irish agitators and English adventurers to take advantage of a unique historical instance of national simplicity. Such is Jonathan's expressed admiration for smartness, that it is impossible to mistake it for anything but the common reverence of humanity for a quality most foreign to the nature of him who admires. In short, only flatter his belief in his own acuteness, and you may get him to take in anything you please—and the harder it is to swallow, the more anxious he is to prove his strength of digestion by forcing it down.

Of course, where there are dupes there will be charlatans of all kinds, but I contend that the greatest American charlatans whom we take for our conventional type are seldom native born. Nor do I forget the class, comparatively small as it is, of cultured Americans who believe that Boston is the intellectual centre of the world. Boston has reflected European culture not unhappily, but that scarcely removes the virtue of simplicity from those who fancy that America has a literature of her own. Nor, on the other hand, do I forget the gross frauds, as gigantic beside the frauds of other lands as Niagara beside a Welsh waterfall. The more gigantic and impudent the fraud, the wider must be the field of those who have been taken in. So obvious is all this, that it is simply amazing how the typical Jonathan has ever found a place even in the airy world of international illusions. With regard to other aspects of his distinctive character, the view is not so clear. He is certainly delightfully self-assertive, and that may be one result of his fundamental credulity. It may also be a reason for our taking him not for what he is, but for what he seems to be. But it must not be forgotten that national credulity is a power in the world, and a very dangerous power. It implies earnestness—earnestness that may take the form of a life and death struggle for a false and deluded cause. If a man, or nation, is simple enough to believe that he has a mission, or that to grow rich is the whole duty of man, or that place-hunting is a worthy form of ambition, or that the welfare of the world depends upon the result of a party struggle, let his friends and neighbours beware. Credulity is almost synonymous with fanaticism, and people may be fanatics, not only in faith or in politics, but in fashion or greed. And on the surface of the seething sea of wild and disordered fancies, demanded by an ever-devouring thirst for what is strange and new, will inevitably float the few who force their personality upon the careless eyes of the outer world. The bird of America is, in truth, less the eagle than the heron, and we have confused the hawk with the quarry on which it preys. No doubt where the herons gather there will the hawks abound. The result is not edifying : but we ought, at least, to be just enough not to hold the feeble gullibility of the typical many responsible for the unveiled voracity of the untypical few.

VII.

PATRICK.

IRELAND is an exceptionally fortunate country. The accepted type of every other nation in the world is but a caricature of the reality, reproducing only its burlesque features, and not always even these. But the world has entered into a vast conspiracy to paint Patrick in oil of roses instead of the ordinary distemper. Erin is the *belle* of the company of nations—made, not to take part in the rough struggle for existence, but to be petted, to laugh and be gay under the most adverse circumstances, to live a rainbow life of April tears and smiles, and to be loved in a sentimental fashion in spite of a few charming faults, and because of many pleasant follies. Patrick has no responsibilities. All he has to do, in society and

in literature, is to be frolicsome even to the point
of head-breaking. The earth's circle is one gigantic
horse-collar for him to grin through. He must
enjoy everything, from a wake to a wedding, and
when he is saddest you may be sure that some
witty word is nearest to the tip of his tongue.
Every school-boy, thanks to Lever and others,
knows the typical Irishman. Barring a few
scoundrels of informers, land agents, and attorneys,
he is the most genial, jovial, hospitable, frank,
easy-going, warm-hearted, amusing, affectionate,
kind, generous, grateful, impulsive of human
beings. It is as impossible to be seriously angry
with Patrick as with a spoiled child, and the sharp
word melts into sympathetic laughter at the
drollery of his gambols. His slight dash of impudence and vulgarity only add a piquancy to this
privileged child of nature. Above all, he overflows
with robust life, and bears the *carte blanche*, which
is the blessed brand of all men who have high
animal spirits, and the courage to let themselves
go. There are people who accuse him of being
untruthful and uncomfortable in his habits. But
he is a charming companion all the same—and,

who looks for perfection in one who is labelled as a
savage?

It once fell to my agreeable lot to arrive by
train at Dublin in company with a typical Saxon.
We took a car together to carry us to our hotel.
After a minute or two my companion burst into a
fit of laughter. As nothing very amusing had
happened since we left our native shores I naturally asked the reason. " What witty fellows even
the carmen are here!" exclaimed the Saxon.
" Fancy a London cabman saying anything like
that!" Now to my certain knowledge the only
words our driver had uttered were, " A car, your
honour?" But he was an Irishman, and whatever
he said was to be laughed at accordingly. My
companion felt himself bound to give our rather
solemn and sour-faced carman credit for all the
humour of tradition. And so it is throughout this
wonderful island. In the matter of wit, for example,
I very much fear that you take away the spirit if
you take away the brogue, just as American
humour evaporates if deprived of its real or
imaginary nasality. It is the accent, not the
matter, which we Cockneys laugh at, not with,

in nine cases out of ten ; and there is no doubt an easy "way" about the "English" of pure Cork or Dublin quality that is itself the germ of laughter. Let everybody consider his experiences calmly. I have known witty Irishmen, but the witty Englishman is not quite unknown, while almost every Irishman is devoid of humour in the true sense of the word. A bull is neither wit nor humour, and collections of Irish wit and humour are nothing more than collections of unconscious bulls. Humour implies a readiness to give and take, not a propensity to answer a word with a blow. The Irishman has always been a duellist by nature, and it is doubtful if a duellist was ever conspicuous for a sense of humour. He cannot bear to think he is being laughed at, and is prone to suspect insult in the lightest word. Indeed, whatever he may have been once upon a time, when we knew nothing about him, the typical Patrick is generally rather a gloomy fellow, though seldom taciturn. And no wonder, considering his climate and his reputation—so galling to any man of spirit—of being little better than a hot-tempered, good-natured buffoon. Then again his leading idea is destructive of geniality. You cannot be in his company four minutes without hearing four terrible words—" The wrongs of Oireland." Be the colour he affects orange or green, the difference lies in nothing but hue. How his countrymen—the most popular and appreciated of races—are trampled on and despised, how his beautiful island is desolate with no one to aid her, how the penurious Saxon will not buy her railways, or transfer Manchester and Windsor to the Bog of Allen, make up the old, old story. Patrick is not only the typical Irishman, but the typical man with a grievance. He hears the " Melodies " running through a symphony of Beethoven. This perpetual self-consciousness is the key to his character. Patrick is not impudent. He is shy—and there is nothing that seems so like impudence as the self-assertion of a man who fancies that his claims are not allowed. He always feels that you are somehow holding him in hostility or ridicule. He even extends his suspicions to his own countrymen. The hatred of an Irishman for every other Irishman throws, in its intensity, the friendship of a Scotchman for every other Scotchman into the shade. Many will be

4

ashamed of being natives of their own country, even while they no less declaim about its wrongs. Patrick can never get it out of his head, falsely enough, that he belongs to an inferior, despised, and rejected nation, and insists upon his view till he almost makes the world think it the true one. He wastes himself in plaintive patriotism, and, with the sensitive temperament of unproductive genius, prefers to be trampled on and unhappy. In short, misery is his joy, for it gives him the right to be a martyr. At the same time, who can deny that the character which clings to defeat has necessarily much that is beautiful? The Irishman can live and die for an idea, all the more triumphantly in proportion as it is hopeless and absurd. That is the highest praise that can be given to any man. Were it not for a quick eye to a bargain, and to getting all he can out of a scramble for places and pensions that entail no work, Patrick, though inclined to be a bore with his eternal wrongs, would have few essentially ignoble qualities. But there is something wrong about everything. If he is warm, he is seldom staunch; if he is staunch, he probably has Scotch blood in him and is not genial. He is not ungrateful, but is gifted with the convenience of a very short memory. He is eminently hospitable, but he is quite as willing to be so at the expense of others as at his own. He receives favours as the payment of a debt, but bestows them as a matter of generosity. Altogether, he is as unintelligible to others as he is to himself; and what is more to the purpose, he wishes to be taken for what he is not, and by dint of eloquence succeeds. He is not frank, for he hides his thoughts and feelings from the prying Saxon as if his very existence were a State secret. Enough has surely been said to show that while the conventional Patrick is a piece of absurd, if delightful burlesque, the true Patrick is an enigma. Let us not, in our wild attempts to do him justice, forget to leave Ireland a few wrongs. No nation can ever be happy except in its own way, and if we were so happy as to succeed in making Ireland happy, Irishmen would be happy no more.

VIII.

ALEXANDER.

"ALEXANDER, King of Macedon," as is well known, "conquered all the world but Scotland." Therefore the Scotchman may fairly call himself the greater Alexander whom friends affectionately, foes enviously, have corrupted into the undignified "Sandy," or even "Sawney." The Scotchman has many foes, and it is only doing him justice to say that he is more than a match for them. It is only the donkey that seeks to devour the thistle.

Englishmen as a rule do not like Scotchmen, and have taken their cue from that arch-Philistine Dr. Johnson, who owes most of his fame to one of the nation he hated and despised—a singular instance of the process of heaping coals of fire.

That cue has led to the present comedy character
—the typical Alexander, as painted by the
Southron. We do not go so far as to picture him,
as Alphonse does, for ever dressed in a kilt and
philabeg. That delusion has died away with the
banishment of the splendid Highlander of our
infancy from the tobacconists' doors, and survives
only on the lyric stage, where Lucy of Lammer-
moor wears the tartan of a Highland clan. But
we do picture him for ever dressed in a character
costume. He is red-haired, tall, with long limbs,
harsh features, and high cheek-bones. His accent
is unmistakable — the typical Alexander always
hails from Aberdeen. The result of a *post-mortem*
examination would be to find "Bawbee" written
where Calais was engraved in the case of our First
Mary. His skull is held to be of abnormal thick-
ness, so that he does not laugh at a joke till three
weeks after it is made, and not even then if the
end of the third week falls on Sunday. He alone
can properly appreciate the poetry of Burns,
seeing that he alone can understand what we are
pleased to term its "Doric." He can, and does,
drink at one sitting as much mountain dew as

would kill a miserable South Briton ; and, above all things, he is true and loyal to his friends to the marrow of his backbone.

I like the feeling that leads a man first to think of his own blood, then of his father's friends, then of his countrymen, and not till afterwards of the rest of the world. It is honest, natural, and wholesome, giving human relationships the proper degrees of precedence. A man who cares for the world before his own kith and kin cannot care very warmly for anybody. A cosmopolitan philanthropist is a man with a very long title, but he is apt to forget that charity has anything to do with home—far less to remember that home is not a bad beginning. But as Scotchmen, even when personal enemies, will stand together like men of one clan, shoulder to shoulder, and will always help one another through thick and thin, it is impossible that we can be quite right in setting down the Scotchman as a cold-blooded animal. Froissart, with his "*perfervidum ingenium Scotorum*," the glowing Scotch nature, was nearer the mark than we. A man who is earnest about everything, must needs be calm, grave, and chary of words. And a Scotchman, the typical Alexander, is desperately earnest about everything—religion, politics, history, the poetry of Burns, the character of his own Queen Mary, the transcendental philosophy, the merits of haggis,—everything that touches Scotland, whether specially or as part of the universe. A Scotch sergeant is in earnest about his drill, a Scotch gardener about his roses, a Scotch schoolboy about his Latin. What is more, he is not, like the German, earnest only for his own or for his work's sake. He is in earnest about human causes, past and present: and if they are hopelessly lost, unpopular, or unpractical, so much the better—he can throw himself into them more keenly. It was not the English Jacobites who survived Culloden, nor is it Englishmen who can launch themselves hotly into historic sympathies, not with any practical political view, but with the enthusiasm of chivalry. I think this to be at the root of the Scottish character—the feeling that makes every Scotchman a knight errant, not to say a Don Quixote, in some cause or other, in spite of a calculating turn proper to a poor country that has always been

more fertile in pounds Scots than in pounds sterling.
I cannot—who can ?—when thinking of Scotland,
keep out of mind the names of Wallace, of Claver-
house, of Walter Scott, of a hundred others who
make a historic round table : and all out of one
small nation not so populous as London alone.
It is impossible to help a little enthusiasm : it is
difficult not to confine oneself to combating *more
Scotorum* the popular belief that the Scot is
"canny," and that there the matter ends. No
wonder that his literature, with one or two great
exceptions, has been barren. Men who throw
themselves into life are seldom writers, and so
much the better for them. They are not apt to
cultivate the graces of manner or style, and they
are apt to be blind to the wit and humour that, as
a rule, spring from the cynicism that finds in all
earnestness something ridiculous. There is no
doubt, consequently, that Scottish humour lacks
wit, facility, and cruelty. Scotchmen have no turn
for epigrams, and cannot understand them. Therefore, while there are no truer friends, there are
many pleasanter companions : and Alexander
himself generally prefers the congenial society of

his own countrymen. He goes out into the great world, for by nature he is not provincial : but it is as an adventurer rather than as a colonist—to fight and overcome rather than to find a home. Of course, too, being very much in earnest, without appreciating the humour of circumstances, he is often stiff and shy—almost always close and reserved. But when he does let himself go, he goes with a vengeance. His hard head alone will carry him farther in a rapid race than other men. It is not only at social gatherings that he will be the last to fall beneath the table. The humour of Scotland lies in pathos—the protest of the warm heart against the rough outside and the hard brain—and what is there more pathetic in the world ?

Such is Alexander. If you are not a fellow-countryman you may find him a dangerous rival—unscrupulous, even, for he is no more the slave of conviction when he is in the right than when he is in the wrong. He will not be ready to extend his sympathies to your hostile point of view : nor will he, because he happens to know you, feel bound to treat you as a friend. But if you need

a friend, servant, or master on whom you can
rely, make him one, and you will not repent to
your dying day—as long as you do not speak ill
of Robert Burns.

IX

DAVID.

DAVID is the long for Taffy, a name intimately associated with Cambrian larceny and Saxon reprisal. It is interesting to observe how types form themselves in infancy. People who make tours in Wales continually complain of the way in which they leave their money upon the road with very little to show for it, either in the way of pleasure or information. The same accident is apt to happen to travellers elsewhere. But no one is taught in childhood that Hans or Alphonse carried off a leg of beef. That unfortunate nursery ballad is, I fear, at the root of the typical David, aided as it is by pictures of market women in steeple hats, and the absurd admission of Lindley Murray that W is a consonant and not

a vowel. From these three elements—an un-
feminine liking for beaver, and a tendency to
consonants and larceny—the Saxon mind has,
after the manner of Cuvier, who could construct
a skeleton from a bone, evolved its typical Cambrian. Mr. Matthew Arnold has but confused
people's minds with his views of the Celtic element in literature and national character. We
accept his views of course, because he ought to
know. But we stand a little puzzled before a
thief who has a genius for music and metaphysics,
and a poet who cannot vocalise. There is no
doubt that between the Welshman and the
Englishman there is a great gulf fixed. We cannot learn his language, he will not learn ours.
While our labourers are busy over pipes and beer,
the Welsh hinds are recreating themselves with
Calvin's five points and Eisteddfodau. Wales
is still a foreign land—foreign in its language, its
thoughts, and its ways. It is rather shameful
that it should be so, and that we should be content to base our international knowledge on
picture-books and primers. In the first place, it
is a little unfair to set down David as a thief *a*

priori. I am not going to enter into a eulogy of all Welsh things, not having the advantage of being born beyond the Marches, and of being therefore, by right and duty, more hopelessly patriotic even than a cockney. But I must suggest, in justice to every Ap Hugh, Ap Morgan, and Ap Howell that a return should be moved for of the number of kid gloves presented to judges of assize, and of the district of Great Britain to which they were forwarded by the glover. That question being disposed of, it only remains to consider the Welshman, secondly as to what he is, and firstly as to what he is supposed to be.

Secondly, then—it is appropriate in dealing with foreign matters to reverse the natural order of things—David is a Celt: a term which, as we use it often, must be assumed to mean something. He is intensely obstinate : he will not believe that the language of the bards is worse or less convenient to him than our own. The Saxons, whose offspring we are proud to be, showed much the same sort of obstinacy after the battle of Hastings. He is singularly good at making a bargain, and looks upon the Englishman as na-

tural food for plunder—a not uncommon error.
He is facile and pliable, in spite of linguistic
obstinacy, and tries to appear all things to all
men, unless you touch his natural pride, when he
flies out into shrill and towering passion. He is
descended from Noah—of course no Englishmen
are—and can trace his pedigree, which is certainly
a distinction. He is born a Dissenter, and talks
Radicalism in his cradle. I do not know that we
have any other special views about Wales and
the Welsh, except that our typical David lives
among mountains, is familiar with goats and
eagles, and exists principally upon butter and
cheese. He is of course as cleanly, pious, moral,
and avaricious as mountaineers traditionally are.
" *Point d'argent point de Suisse*"—no money, no
Welshman. Firstly, then—there is no doubt but
that there are mountains in Wales, though goats
and eagles are becoming obsolete. But as to his
race, has it never struck those who have studied
such matters at first, and not at second hand,
that he is very different from other so-called Celtic
races—the Irishman, the Manxman, and the Highlander? The fact is, he has in him a large ad-

mixture of the blood that conquered, not of that which was subdued. The language of Wales contains a Roman element, not the result of literary importation, but bound up with common words. I must content myself with assertion, not having inclination to set out a philological vocabulary. There is a popular tradition that the Romans drove the Britons into the Western mountains. So they did, and followed them, and there they remained when the legions were withdrawn. There are as many " Ponts " and " Strads " and so forth in Wales as in the kingdoms of Kent or Northumberland, and you will look for tokens of Danes and Saxons in vain. Now where the Celtic race has come in conflict with the Latin race, the latter has triumphed : and therefore, I take it, the Welshman has as much in him of the Roman as of the savage whom the Roman found. He is not brilliant like the true Celt : he has the plodding industry of the masters of the world. Like them he is stubborn, and stands on his dignity of race and person. He is grave and practical in his humour, with a tendency to rhetoric rather than poetry, which comes from the heart

and not the brain. He has a stupidity peculiarly
his own. It is not the gross stupidity peculiar
to the Saxon boor, but it is that stupidity which
will not permit a man to understand that he is a
provincial. A Welshman cannot get it out of his
head that he belongs to the greatest, most important, most central race in the world. To him
London is provincial : he is in the world's capital
when he is in his own country town. His an-
cestors the Romans thought the same—the only
difference being that they managed to impress
their belief upon other people too. But, as the
modern Welshman cannot do that, he makes up
for it by acting as though he had done so. To
sneer or speak lightly of Wales, to say that
Welshwomen are not the most beautiful in the
world, that Welsh music is not the most musical,
Welsh poetry not the most poetic, Welsh mutton
not the tenderest—as in fact it is—Welsh scenery
not the loveliest, Welshmen not the cleverest,
Wales not the most wonderful, is to elicit a cry
of agony. He bristles with prickles, and if you
touch his country you touch him. Being essentially provincial, he is, if not necessarily narrow-

minded, devoid of the urbane humour, and there
is no doubt that the typical Welshman is rather
prone to be suspicious of his friend. Mutual trust
and confidence is not the result of a life bound up
with country villages. It comes from wide knowledge of the world, and the world the true Welshman does not care to know. He is not at home
out of Wales, and carries his native valley on his
back wherever he goes. But, if he does not trust
his friends, how shall he trust the stranger?
" *Timeo Danaos et dona ferentes*"—I fear the
Saxons, though they bring us gifts—was put into
the mouth of the Trojan, a common ancestor of
Wales and Rome. If we think he wants to cheat
us, he thinks we want to cheat him. That is the
true philosophy of the story of the mutton bone,
which, after all, cuts both ways. The Welshman
would be like the Roman if he could. He is as
philosophic, as literary, as obstinate, as patriotic,
as dignified in all his tendencies. If England were
not England, he would no doubt condescend to be
less provincial and more imperial. As things are,
he might fairly accept facts, and, by becoming
more English, become yet more Roman.

5

X.

JOHN.

THERE is, of course, nothing remarkable in the fact that Alphonse, Hans, David, Patrick, and the rest of them draw strange pictures of John. But it is not a little peculiar to notice how curiously careful he is to misrepresent and caricature himself. One would think us to be the shyest country in the world, or else the most cunning, to see the way in which we throw about printer's ink, as if we were all literary cuttle fish, to blind the eyes of our friends and foes. We delight in looking at ourselves neither as we are, nor as we wish to be. We laugh at the red-haired, long-whiskered and sloping-shouldered imbecile, with the long teeth, who stands in the *Charivari* for " *Sir Brown, Milord Anglais*," but we do not treat ourselves with any greater fidelity to nature. "The English painted by themselves" make up not a portrait gallery, but a museum of monstrosities. I do not quarrel with my countrymen for claiming a monopoly in the name of John. I have known a few Englishmen who were christened otherwise, but on the whole I suppose there are few families in which the name of John is entirely unknown. It certainly has a curious affinity to the name of Smith, which, oddly enough, has no correlative among the Latin races. But I do quarrel with our choosing to add to the name of John the surname of Bull. That very unhappy thought of Arbuthnot has been productive of evil that is something more than merely pictorial. John Bull is even still the conventional Englishman. I am ashamed to recall his characteristics. He is a bumpkin, a clown, and a boor. His gross, stupid animalism speaks in every feature and every limb. He is bloated, red-faced, and weighs twenty stone. He not only eats beef till he becomes bovine, but drinks beer till he thinks beer, and habitually wears top-boots in the drawing-room. The comic journals of a country are generally supposed to reveal its true characteristics,

grotesquely, if you will, in manner, but faithfully
in spirit. Therefore we know what will be thought
of us in future times. Not only so, but we have
taken a traditional kind of shame-faced pride in
the lout whom we dare to call a likeness of ourselves. His very girth we account a grace, and
distrust any but the brutal virtues. No wonder
that other nations take us at our word and hold us
for the *bêtes* that we claim to be.

We are most of us still proud of being Englishmen. But the justice of our pride lies in the fact
that we are in every respect unlike John Bull. If
we weighed twenty stone, could we be the best
horsemen to be found after the Tartars? Could
we lead the way up the Alps, and win the race to
the North Pole? Were we bumpkins, clowns, or
boors, could we be the most aristocratic of all
nations, next to none—with a passion for comfort,
which is only another name for the highest sort of
refinement, and a fastidiousness which is in itself
a mark of a thin moral skin? Were we mere
broad shouldered but narrow-minded Minotauri—
but science, commerce, literature, stand wonder-
struck at the idea. No: we have our faults, but

these are not of them. On the contrary, our faults are precisely the reverse of those of John Bull. I cannot say the same of merits, for the conventional John Bull has no merits either in body or soul, to judge from his personal appearance, except a look of apoplectic health and a purse well provided.

The main characteristic of the true John—not Bull—is that he has no common characteristic at all. There is no typical Englishman. It is easy to distinguish an Englishman at a glance from a foreigner, but that is the result of his individuality, not of his nationality. Of a foreigner whose individual character is strongly pronounced physically, and who shaves sufficient to show it, we invariably say "He looks like an Englishman." This is what Goldsmith meant by calling us a nation of humorists. Our minds and thoughts vary as much as our climate, and our bodies as much as our minds. Hence almost every man has tastes and ways that render him strange to his fellow-countrymen. It has been observed of a French crowd at a great spectacle, that every man has the same impulse at the same moment, and expresses it in the same word. If one exclaims "*magnifique*,"

all exclaim "*magnifique*," if one *superbe*, all *superbe*
—not taking it up one from another, but in chorus,
as if there were but one voice to the crowd. In
England, though we are slaves to fashion, out of
a self-consciousness of our individual eccentricities,
unanimous impulse is almost unknown. It is not
we who experience panics in battle : and if fire
panics are not unknown that is because we are
human beings, and women are in such cases
mingled with men. This characteristic takes a
twofold form. Some it makes selfishly intolerant :
others it throws into eager sympathy with the
thoughts and workings of other minds. We are the
most bigoted of bigots when we are sure we are
right : the most timid of sceptics when we are
not sure that we are not wrong. In fact, an open-
minded Englishman, as most Englishmen are when
young, is seldom quite sure of his own mind.
There are so many ideas floating round him, with
somebody to say something for all of them, that
he must perforce take refuge either in bigotry or
scepticism. So he temporises, and contents himself in practice with what is certain, finding in
tradition and in conventional custom a substitute

for the trouble of having to make up his opinions.
Therefore the Englishman is essentially a "respectable " man. He leads the common life of those
immediately about him, not so much because he
thinks it right, as because he has to choose between
the comfort he loves and the very uncomfortable
strain of perpetual self-assertion. At last submis-
sion becomes a habit, and he looks upon displays
of eccentricity as rather weak-minded—as, in fact,
they mostly are. The result of this is, however, a
greater amount of unconscious hypocrisy than can
well be found in any other part of the world. You
can scarcely ever tell what an Englishman really
is from what he seems. Indeed, so much does
prudent common sense teach him to subdue his real
nature that it is not unfair to judge by contraries.
We are always astonishing one another, and we
seldom know even our dearest friends till chance
breaks down the icy barriers of habit, and shows
that Englishmen are men. We hold ourselves back
till the occasion comes : and that I take to be the
cause of our enduring strength, which we do not
exhaust by frittering it away in hourly outlettings.
There is no doubt, however, that while strength is

increased by repression, readiness of emotion is destroyed. The emotional artistic temperament is produced by indulgence and cultivation. And so among us genius must, in order to be recognised, dare more, and dare to be more eccentric, than in countries where singularity is not looked upon as a sign of weakness. So, also, the pursuit of the arts most allied to genius is still, among us, looked upon as a mark of singularity like a *bizarre* costume. Like other people we cannot be : and so to be like other people is, of course, our aim. Failure makes us perhaps a little awkward, and gives us a reputation for want of sympathy which we are really the last to deserve. "As many men, so many minds." He differs essentially from Alphonse by keeping up the traditions and the spirit of courtesy. From Hans by caring more for the results of work than for work as in itself an end, from Giuseppe by readiness to spend, from Juan by richness of nature, from Ali by originality, from Jonathan by incredulity, from Alexander by love of comfort, from Patrick by a sense of humour, from David by being able to appreciate the fact that there are other people besides himself in the world. But he almost resembles the Welshman in his belief that his own country is the centre and capital of the universe, the Scotchman in his want of expansion, the Irishman in his physical activity, the American in his love of all things new, the Oriental in his slavery to fashion, the Spaniard in superstition, the Italian in versatility, the Frenchman in being the biped which is called man. Enough, however, has been said to show that, if not the pleasantest nation in the world we are certainly the most remarkable : and that with all our faults we are not, in our type, a horde of boors or bulls.

FLORA AND FAUNA OF LONDON.

FLORA AND FAUNA OF LONDON.

I.

AMONG THE "GODS."

THE Romans had their first-class gods and their second-class gods. So we, who imitate them in so many things, have our "gods" who pay sixpence for the privilege of Olympus and our gods who pay half-a-crown for a more exalted elysium, To-night, while we have five sixpences, we mount into the loftier region. We are at the Opera.

Your ladyship thinks that the opera is made for such as you? You, bouquet-bearing mortal that you are, are unaware, even with the aid of a *lorgnette*, of the presence of those who sit in a world above the chandelier. You do not even

glance up at them : but they, in a double sense,
look down on you.   They are a conceited race,
though yours is the highway of the crush-room,
theirs the byeway of the back-stairs.   They are
the true *fanatici per la musica,* not you, who put
on kid gloves, and arrive comfortably in your
carriage at the end of the overture.   Music, like
war, is no kid-gloved affair with them.   Long
before the doors open they crowd through the
narrow entrance and up the steep and stuffy
stairs that lead to their paradise : for a good
three-quarters of an hour they rejoice in purga-
tory and breathe without oxygen.   And then the
rush when the bolts are drawn !   You would not
do as much were the ghost of Pasta to return to
take a last " farewell."   Well may they think it is
to them that singers sing.   Well may they expect
some gratitude for going through so much to hear
what, alas, often in these degenerate days turns
out to be so little.   The sixpenny gods are easily
pleased.   Not so these.   Hard seats and a hard
fight to gain them do not conduce to the languid
satisfaction proper to the boxes.   They talk
between the acts : and then, how they criticize !

They seem to enjoy being displeased better than being pleased. But their opinion, unless, like Mezzofanti, one knows all the languages in the world, is hard to gather. There is a lot of German clerks, who have come, a volunteer *claque*, to applaud Fraülein Katzkorff for old recollections' sake of Vienna or Berlin. There sit Monsieur and Madame, with the little Alphonse and the little Alphonsine, for all the world as if they were at home in the Porte St. Martin. These wear a solemn air—to them, the theatre is a function : to the Germans, a recreation only lacking beer. By your side sits the twin brother of the Garibaldino who, after a long day's march to Milan, spent his only *sou*, not in maccaroni, but in a thousandth hearing of " La Sonnambula." This one, for to-night's sake, will starve for seven days. Be you of what country you may, you will find fellow countrymen — and fellow country-women, too — with whom to fraternise, where liberty, equality, and fraternity, politically and socially, are the order of the day. It reeks with Bohemian air. The English element is less interesting. It is indeed less important, for you may sometimes

sit out a whole performance, and not hear an
English word. Of course there are specimens
of that indescribable order of beings who are
found everywhere, and resemble nothing but flies
in amber. But most are able to point out with
pride some member of the chorus or orchestra
whom somebody that they know knows: or
others of whom are themselves struggling music
teachers or musicians, who have been extrava-
gant enough to spend upon the art that starves
them the hard earnings of half a day: and these
are the devoutest listeners of all. But only avoid
the old gentleman who "Saw Grisi, sir, in her best
days," the middle-aged gentleman who is always
losing his place in his *libretto*, and the bandsman
off duty who drums with his heels to the music,
and you will do very well. You can talk when
you want to talk, and listen when you want to
listen. What *habitué* of the stalls can do more?

It was among the gods that De Quincey
dreamed opium dreams, listening in the intervals
of Grassini's song to the music of the Italian
language talked by Italian women. Perhaps
opium had something to do with this last plea-

sure : for Italian talked by Italians is not the most musical tongue in the world. Let us, therefore, return to the músic of the English language, as it is spoken by English women in the stalls. It may disturb the performance : but for that I am too patriotic to care—and, as I have already said, the seats of the gods are very hard.

II.

AMONG THE STARS.

WHEN a troupe of travelling musicians—I trust
that *prime donne* will pardon the use of a title
that they share with the stars of fairs—arrives in
a country town, it is received with some amount
of awe. Such names as Corbacchione and Barba-
gianni, hitherto known only from the musical
advertisements and criticisms of *The Times*, become
no longer symbols for abstract ideas, but names
of audible men and women. No lady of well
regulated mind would receive them as guests : nor
does she so make any unfitting sacrifice of charity
in many cases. But she, especially if she is herself
an amateur, feels a half sympathetic interest,

tinged, of course, with the sweet flavour of impro-
priety in them and their ways. The waiters and
chambermaids of the Royal Hotel, where they
take up their quarters, are of course in the rela-
tion of the valet to the hero. But in the concert-
room what mysterious glory exhales from the
brow of the finely if not very tastefully dressed
lady who, every time she opens her lips, gains
for herself or her farmer — for *prime donne* are
generally bought or hired—the price of a farm !
It is of course presumed, even by those who know
nothing of her but what they hear—in a double
sense—that she is a woman like other women.
But somehow people who read novels, which now
very considerably run into art matters, take her
as the representative of a great, poetic, and
romantic career. If they go to the theatre, they
see her fed with applause, and turned into a living
statue of Flora. If they meet her in society,
they are themselves a little disappointed with her
style, but she is so surrounded by courtiers that
they are ashamed to own their disappointment
even to themselves. If they read the papers, they
read of her as if she were something altogether

divine. Even her faults of style or intonation are
discussed with more solemnity than if they were
those of a statesman. The Chinese used to think
that there was no world outside their great wall.
So does the world of amateurs consider themselves
the outside Tartars, who look with longing eyes
upon the flowery fields, as they deem them, of
song.

Far different to the amateur is the star. She
has been brought up to the trade, and knows its
traditions. I am talking, remember, of an imaginary orb. She is the vocal representative of a
glorious art — an art more soul-absorbing than
painting or poetry. Once, perhaps, she possessed
a little private enthusiasm of her own. Talk to
her now, and you will not be called upon to strain
your transcendental faculties. She will be best
pleased if you confine yourself to praise of herself,
and finding fault with her rivals. Not that she is
naturally jealous or more than ordinarily vain, but
business is business, and all is fair in trade as in
war. Besides, self-belief is a habit very easy to
acquire. She knows beforehand from what part of
the house the applause begins, where there is to be

an encore. She will even keep her show song back to produce at the right moment. In that case, if there is no encore, she will take one, all the same. Her happiness in many of the bouquets thrown to her must be small: she likes money, and bills run high in Covent Garden, or in the *marché aux fleurs.* If, as is commonly the case, she has a fair share of common sense, she knows that the people who crowd round her like moths round a candle, are swept together by reflected vanity, or by interest. They wish either to be patrons or protégés. She finds no pleasure in delighting the few who run after her voice, or her skill, or the thousands who admire her because it is the fashion. Her aspiration is a villa on the lake of Como, with nothing to do: and when that is fulfilled, she continues to take an interest not in art, but in gossip about its *personnel.* She is, generally speaking, a very good sort of woman, in spite of what people may say: but to entwine a *prima donna* with poetic wreaths, is about as absurd as to treat a cobbler as though, because his boots were the fashion, he ought to have a public statue. He does the best for himself, and so does she.

The true enthusiast, the true artist in song seldom becomes a star. She—perhaps—has the reward ascribed to virtue. A star may be glorious in itself, it is true : but far more often it shines with reflected light, and fashion is its sun.

III.

AMONG THE GRASSHOPPERS.

THE noisy insect into which Aurora turned Tithonus sings, says the legend, not because it
chooses to sing, or because its song is sweet. It
sings because sing it must, unceasingly, until
it bursts in pieces with the perpetual strife to sing.
The latest addition to the live stock of a certain
horrible old crone who traded upon the beggary of
children must have had the blood royal of Tithonus
in her veins. I will come to her presently. Mean-
while, enough and to spare has been said of musical
miseries. All the world asserts that barrel organs,
ballad singers, *pifferari*, and errant bands are a
nuisance of the order intolerable. All the world
thinks—though it dares not own—that "benefit"
concerts, which benefit nobody ; "recitals," which
mean a solo performance upon a private trumpet,

musical mornings, musical evenings, musical parties, musical festivals, musical amateurs, connoisseurs, professors, critics, and charlatans, are but the development of a street nuisance upon a drawing-room-scale. Talk of the pleasures of music! Its pleasures are like the visits of angels : its miseries are as much with us as are the poor. Hateful though it may be in politics, in Art there is no principle so sound as that of " Liberty, Fraternity, Equality "—above all, " Equality," the key-note of the three-hued symphony. The uncritical audience of yonder Savoyard—the earliest instructor that our princess of *cicalæ* knew in our earliest years—is listening for music's, not for fashion's sake. Beppo himself is playing for a no more ignoble cause than that which evokes the song of his more aristocratic compatriots in the opera house across the way. What is the essential difference between the two? Every now and then an artist astonishes the stage by showing that the lyric soul may have room for other things than a weekly salary. But the street also is not without its artists, who, if not so great in reputation, are in spirit equally true. The owners of fastidious ears will doubtless prick them with wonder. But I say boldly that of the few artists, worthy of this much-abused name, to whom I have listened with a feeling that they stood in a world wherein criticism becomes only another name for impertinence, two hold a foremost and yet a common rank. One sang for glory and for glory's gilded frame: the other for halfpence without any glory that needed gilding. But, with both, it was Art that stood above and in front of all.

Poor child! It is hard to call what in others is termed a graceful accomplishment, idle vagabondism in you. Is not vagabondism the very sign and seal of the artist nature—and in what way is it more blameable to wander from London street to London street than from Vienna to San Francisco? Perhaps you have gipsy blood in your veins, and are a wanderer by right of race as well as by right of nature. Has it ever been remarked that nine street singers out of ten have black hair and tawny skins? This girl's hair is as black as ebony, and her complexion— but that is not so easy to tell. We are standing not far from Bow Street, where the law is investigating the rights and wrongs of the wretched

Beppo.   Beppo also is a grasshopper ; and his
compelling fate is the *padrone*.   But listen.   "That
is not a bad voice," you say : "who knows what it
might have been with training ? "   " Not a bad
voice," indeed !   It is a glorious voice : and, as to
training, it is not so much instruction as oppor-
tunity that is needed to make it that of a Sontag
or of a Malibran.   She is the very queen of grasshoppers, though her robes are tatters.   Her compelling fate is Nature.   Her blind soul does not
know it : but it is not for the sake of stray
halfpence that she wastes her treasure.   As for
what her life is when she is not singing vile words
to viler tunes—well, it may be no worse than if she
had been an opera queen—a queen, not among
grasshoppers, but among dragon-flies.   There are
countries in which she would have been caught
and trained.   London is not one of them.   We are
ants : we have no fellow-feeling with grasshoppers.
The rattle of our streets drowns all but the song of
the lark, who sings to the skies alone, though he
cannot see them through his prison-bars, and to
the sod which has to stand to him for the whole
green world.   In a few months' time her song will

be cracked and hoarse, with London rain and wind and fog, and worse things even than these. In a year, where will she be? Perhaps she will have followed her dead voice : perhaps she in her turn will have turned *impresaria* of beggar children. In either case, Nature has mocked Art by bestowing one of the sweetest and most gracious of her gifts in vain. But mere street ballad singer, street beggar as she is, and will have been to the end, she will have done more than most of us. Scarcely one in a generation bears the fruit of song—how few even bear the blossom ! Is it to enter into the forbidden world to fancy that Nature gives not one of her gifts in vain—that song, though existing but in promise, is the true soul of a born artist, be she *prima donna assoluta* or be she but *prima cicala*—first of grasshoppers ? In that case, when you hear a street singer, with her hoarse, cracked voice, think, not of the torment to your eyes and ears, but of what she might have been. This is no fable, though it set out with one. The grasshopper, with her one strident note, is not the less a singer because she is slain by song.

IV.

AMONG THE STALKS.

ONCE upon a time, a long time ago, the making of nosegays was one of the simplest arts known in Eden or anywhere. Nothing was needed but to pull a handful of cowslips from the nearest meadow, or of anemones from the hedge, and to bind the stalks together with a tendril of the wild clematis, which some call traveller's joy. But we, in our cleverness, have changed all that, even though we may not go quite so far as Sganarelle, who, on his own authority, transferred the heart from the left to the right side. It is probable that the hearts of men and women are where they have always been, and that all who have been children have in them still a quiet place for wild flowers. Only the ingenuity of florists has raised a wire fence between the heart and the blossoms that nature bids it fear.

Bouquets—I cannot call them by so unprimitive a word as nosegays—are now manufactured by wiring together into a symmetrical pattern gaudy petals and heavily-fragrant blossoms, while the stalks are thrown away as an incumbrance to a kid-gloved hand. Covent Garden, wherein we are now standing, is the metropolis of these bouquets. But it is something more. It is itself a bouquet. Here are not only the bright petals of flowers but of flower buyers, who lay down gold with hands that would grace Aurora. Here is not only the scent of blossoms, but the perfume of the memories of the Piazza's flowery days. And here also are the stalks that are thrown away.

I do not mean only the stalks of cabbages, the empty pods of peas and beans, and the thorny shreds through which roses drew their life from the soil before they had been transplanted from the bed to the bouquet. Everybody knows Covent Garden at noon. Then it is in blossom. Most people know it, or rather have seen it—for the knowledge of the early sparrow is shared with few

—in the fresh, cold dawn, when market gardens, like very woods of Birnam, march into London as into another Dunsinane, bearing with them their diamonds of country dew. Even the tenants of front rooms in the innumerable inns overlooking the waggons do not complain of a clamour that wakes them only to send them to sleep again. But it is not always noon or dawn. Covent Garden is not always the Convent Garden—a city sanctuary where rose abbesses rule over lily and violet nuns—a floral cathedral in the midst of an unfragrant world. There are some few hours of night when even Covent Garden tries to sleep, and when the stalks forget in dreams that they have forgotten to bear flowers.

She who sleeps in a torpor of exhaustion, among baskets for pillows, and with the night air for her covering, is a stalk. She is a stalk that once upon a time—how long ago it seems!—drew from the soil the sap that feeds the flowers. Her pillow then was that of health and growth: her coverlet the air of home. There is more in his doctrine of the metamorphosis of plants than even Göthe, its inventor, knew. He called the development of flowers from roots a progression not from imperfection to perfection, but from strength to weakness—from health to disease. First comes the stem or stalk, firmly planted in the soil of earth, which, for flowers, never ceases to be the soil of Eden. Then the branches spread themselves as if to grasp all the outer air. Then, as the sap grows weaker, come the leaves, to shield the plant from the very sunshine towards which it strains. Lastly, break out the flowers, to end either in seasonable fruit, or in—a bouquet, to be gathered, carried through a ball-room or crush-room, caressed, and tossed aside. So from above that woman, who sleeps, let us hope, and dreams, we may be sure, among the refuse both of roses and of cabbages, may we, in like fashion, draw aside the transparent foliage of her dreams, until the constable's bull's-eye flashes upon her from the piazza and warns her that there is no rest for such as she. The artificially wired blossoms have long ago—or lately, what matters it?—dropped away from their wires. The leaves are no longer there either to shelter from the scorching sun, or to receive the freshness of the rain. The root is torn

up from the soil. This is the process of the manu-
facture of a bouquet. And what has become of the
stalk? It lies here among the rejected waste of
Covent Garden, whence Waterloo-bridge is not far.
It may be thither that the glare of the lantern will
bid this human stalk "move on," and our great
river will carry to the sea the refuse of a garland.
Yes, Covent Garden is in every sense a bouquet.
Its market-place is the very type of those who
snatch among its shreds and refuse such dreams as
may come between the close of a black night and
the coming of a misty morning, bringing with it
the cartwheels of the busy day to crush into the
mire the stalks that had once borne blossoms in·
the fields.

V.

AMONG THE BLIGHT.

IN my wanderings to discover the saddest possible thing that the world contains, I thought it fit to confine myself in the first place to my own country, and in the second place to London, simply because England contains London, and London contains all things. But once lodged here, the search became more difficult. It was hard to choose between the arches of the river-bridges, the prisons, and the drawing-rooms. But at last I came to a door, on the lintel of which suddenly flashed out the word "*Eureka.*" It was a hospital for sick children. I am not going to describe its locality or its organisation. It is a bad principle to satisfy at second hand the curiosity that ought to satisfy itself by becoming charity. But does any one remember how sometimes, when the morning is young and the glass is at " set fair," when there is not a cloud in the sky, and the day lives in hope instead of memory, there comes an obscure darkness which is not mist, or the presage of fertilizing rain? The country people call it "blight," and its whole purpose is to destroy. It is one of those instruments of nature that seem to us to be wholly malevolent. Thunderstorms, though they devastate the fields, have their benevolent as well as their glorious side. But blight, though less widely destructive, is both wholly hideous and wholly evil.

Here is one little child, who has seen eight years, of which each has been a winter. It was not born a cripple. Had its mother been a negress or a Sioux, its limbs would be as straight and its life as strong as that of any of its brothers and sisters who are brought up in the forcing houses that we call nurseries. Its sin has been that it was born in the most civilized city in the world, over one half of which the sun shines gloriously, while over the other half—or is it more than half?— spreads the stormless blight of bad nursing, worse

food, and worst air. That day should end in night,
is fitting : but that the twelve hours of midnight
should strike with the first hour of dawn—have I
not discovered the saddest sight in the world?
Look at those small rays of hands, which nature
made to grow till they could wield the axe or
hammer, and which our second nature might have
taught even to hold the pen. Look how the
features, which were meant to smile till strong life
made them grave, have in eight years of childhood
only learned to express the patience of hopeless
resignation. Look at the limbs that ought to be
stretching themselves out into the air of the fields,
even if only into the fields of St. Martin or of
St. Giles, striving with those of other children, not
which can run the fastest, but which can lie most
still. And then turn from bed to bed, and think
upon the suffering of innocence—a paradox which
is purely human : and of how that suffering falls
hardest upon the children of the poor, who most of
all need the strength of shoulders and of hands.
These cannot compensate for defects of body by
the triumph of mind over the strength or over the
weakness of matter. Here is another child that

has undergone one of the most terrible operations of surgery : another whose life is only being prolonged for a few short months in order that it may die of disease instead of starvation—each, unable to reach out his hands, reaching out his eyes, with touching greed, for the wooden horse, or piece of coloured card, that stands to him for so much of the universe as is not contained in the words deformity and disease.　Even if any one of these lives to be old in years, he will never have known what it is to be young.　The day of life is reversed, and the infant is born into the seventh age.

And yet, in the midst of premature disease, and in the face of death, who comes hand in hand with birth, one sees through the veil of blight the outline of something which is not that of the destroying angel.　We are in London, and in the midst of roaring streets and stifling lanes.　But through the shadow the passing wheels are unheard, and the reek of the close courts is filtered and purified.　If the blight has forced its way within the walls—if the sunshine and the showers are left without, it has called forth the care of human gardeners who neglect too much the wild and healthy blossoms that are exposed to the scorching of the sun and the rush of the storm. If only there could be a children's hospital for all—for the strong plants as well as for the weak, for lives that need to be restrained, as well as for lives that need to be saved! These sad, sickly, but resigned faces, these feeble but patient hands, these minds closed to all the evils of the world but one, all these children who are, save in innocence, all things that children ought not to be, have learned in childhood the fiction, better than truth, that the earth is ruled by kindness, and the truth that death is not a thing to be feared. No : sad, unaccountable as it is that blight should be carried by the breeze that carries day-break, I must wander farther before I can say that I have found the saddest thing in the world.

VI.

AMONG THE CATERPILLARS.

ROSE is the best possible tint for spectacle glasses. They render the poverty of other people invisible, they give pleasant names to unpleasant things, they call a spade anything except what it really is, and by a severe process of translation, have turned Grub Street, St. Mary Axe, into Milton Street, in the same unsavoury locality. But no spectacles however exquisitely tinted, can turn a caterpillar into a butterfly. If they could, how glorious the world would be—and how full the Abbey would be of tombs! But, as things are, the ways that Goldsmith trod are trodden still. We often bandy compliments about Grub Street being no more : but we do not change things by changing names.

It is difficult to be geographically exact. But the modern Grub Street—I like to use the good old name ; it is redolent of poetry and of pathos, in spite of its unpoetic sound—may be taken to lie within a circle of byeways that radiate from Covent Garden as a centre. To you, that quarter of London, our mother of mystery, is but the land of music, of theatres, of late suppers, and of early peas. This is because you keep to the highways of the world. Turn down this narrow winding alley and enter this untempting tavern door. You must not mind if your rose-coloured lenses are dimmed by the smoke of shag, by the fumes of malt, and by clouds of spirituous steam. You will see typified for you many another tavern in many another winding alley within the cast of a stone. You have passed the threshold of a land wherein all things are upside down.

The " Wormwood Arms "—that will pass for a generic name—is a dilapidated house of the days of the earlier Georges. It possesses historic and classic traditions. But it is a failure now : and it is here that gather the Failures—men who have taken to the underground ways of literature because they were, or more probably thought themselves, unfitted for safer and easier ways. Has it ever occurred to those who study the secret of success to fathom the philosophy of failure? They will here find a whole hospital of cases. These men are the caterpillars, who, with the making of butterflies in them, will never become butterflies. Each lacks some one little thing—each has some slight moral or mental twist that keeps him under the water of life, on which many a worse man triumphantly swims. There is nothing so terribly bitter as to be *almost* good enough—to miss the centre by just the fraction of a line when to miss, even by the fraction of a line, is to fail. Make up to Jack Hacker, there in the corner which is by prescription. From one, learn all. He is a wit and a scholar—nay, he is a gentleman, though a rough one, still in spite of loud talk and a suit of clothes that would disgrace a militiaman in mufti. He keeps himself from starving by his pen, not even himself knows how, but beyond that he has always failed. He was the son of a man of fortune, even of rank, and was intended for the Church or the Bar. But he was plucked for his

Little-go, though capable of examining for honours. He never played billiards without just missing the stroke that was to ensure him victory. He never backed a horse but it was beaten by a neck. He never wrote a line, good or bad, but he was told that "it would very nearly do." He missed winning a wife—a girl whom he treasures still in his heart in a secret and shamefaced way—just because, as ill-luck would have it, he in those days the soberest of men, drank one glass too much— he has often done so since, poor fellow!—on the eve of his wedding day. By just one second he invariably loses the post or the train. When things were at their worst he tried the coward's escape from trouble, but even in this he failed. So he joined his fellows at the "Wormwood Arms," in Weevil Lane. It is his substitute for home. He and they have become what the prosperous and successful ignore, and have to accept failure as they best may, some defiantly, some querulously, some desperately, and some patiently. Even when, by chance, the lane turns at last, and some cater-pillar does happen to become a butterfly, the trans-formation of their comrade does but little good

to them.   He spreads his wings in the sunlight of
editorial air, and forgets too often that he too was
once even as these.   And then his old friends unite
to abuse him, and all his works and ways.   For
their creed is that to fail is to deserve to succeed :
that to succeed in to deserve to fail.   And sometimes the creed is not wrong, sad and bitter though
it may be.

Go into the libraries of our great museums,
and you will see to what a last pass the drudgery
of hack work still brings men—aye, and women
too.   It is simply terrible to see there the haggard
faces, the wasted frames, the hopeless aspect of
many who cannot even reach to the consolations
of the "Wormwood Arms."   No : Grub Street is
not swept away.   It has but changed its name
and widened its area—that is all.

VII.

AMONG THE MOLLUSCS.

DURING the months which contain the letter "R" in their names, I cannot help regarding that interesting creature which is "neither fish, nor flesh, nor good red herring"—the oyster—as the true height of human happiness. I do not mean as an edible, delicious though he is, but as a type. He has a home, far down among the depths whence the wildest tempest is not strong enough to tear him. The waves roll and the winds roar over him, but he is great in his safe and self-sufficing calm. He has but to open his mouth or door and all his wants are supplied. He is an eminently respectable and respected mollusc—the cynic of the sea. He is destined to be valued even above his merits—at the rate of something like half-a-crown a dozen. He is protected both by nature and by legislation: and, when his career is run, or rather stood out, it is to receive, by way of epitaph, the thanks of some grateful fellow-creature for benefits conferred. No wonder, with so sublime an example daily before our eyes, we do our best to emulate our fellow-native. Do we not turn London into a gigantic oyster bed, and cling to its bricks with all the force of our fibres, hiding as much as possible from the risks of sunshine and of air? Do we not bring comforts to our mouths and doors, so that we have but to open them to take them in? Are we not proud of being the cynics of the world's sea? Do not we also value ourselves even at something more than half-a-crown for a dozen of us? And if we can in the course of our life find one grateful fellow-creature who will speak kind words of us after we are gone, do we not also feel that we have not lived in vain?

But, alas for human, if not for molluscous, perversity! One fine morning the letter "R" slips away from the month's name. It leaves a little hole through which pierces a ray of what is to be the summer sun, fragant with the breeze that has swept over the flowers of the sea. It is May, the mother of June. The swallow has long since taken his passage for the sources of the Nile. We open our shells just a little, and our rock does not feel quite so soft as usual. The scent of the first strawberry, the first stray taste of hay that some imaginative people fancy they can catch even in the Temple, the chance sight of such words as Baden, Bruges, Prague, in the tables of an obsolete continental Bradshaw, seem suddenly to set the heart of the stork or swallow beating in the bivalve. Daily, humdrum life, though it may breed the pearl, seems to prove that pearl-breeding is, after all, a disease. Every well-regulated mind is seized with a glorious desire to be a little insane. The mollusc develops more swiftly than the laws of Darwin, and even amid the depths of the rocks feels the budding of inconsistent wings.

There are paradoxical people who hold that London is pleasantest when emptiest. The streets are no longer crowded. Cabs and peaches are cheap and plentiful. To dine *en ville* becomes almost comfortable. The drop in the ocean

expands into an almost recognisable wave. On
the other hand, travelling is more full of trouble
than ever. Most things and places are more beautiful in the imagination than in their reality, while
the history of a tour is little more than a record
of trivial troubles and very untrivial bills. Were
it not for the longing that makes absence of discomfort the greatest discomfort of all, how happy
would the imprisoned oyster be! It is something
more than the shame of not being like her neighbours that induces even a lady to draw down her
front blinds and live in her back rooms rather than
confess that she has not gone out of town, or to
make the payment of a morning call in summer
time the offering of an insult. It is that the soul
of the oyster has outgrown its shell. It would
fain be for a while one of those homeless, ram-
bling, Bohemian fishes that have no rock and breed
no pearls, but are carried by wandering currents
into the shark's jaws. We may bivalvise ourselves
as hard as we please, but the instinctive love of
novelty is too strong—we are still passage birds in
soul. When the grass seems to peep between the
flags of Regent-street and the clover in Pall-mall,

then the billowy clouds insist not on suggesting a rainfall, but the Alps—the Thames, on talking of the sea whereto it flows. Home-sickness is bad enough, but nothing can compare with the sickness of the swallow that pines to flee away. What then must be the desire of the oyster? It must transcend that of the moth for the star.

Yes—if the lot of the mollusc be happier than that of the bird, then the sea anemone must be more blessed than the oyster, more blessed than the anemone the stone to which it clings. But though the swallow loves its nest, and can fly to it as straight as an arrow, over the mazes of a thousand miles, he leaves it when instinct makes home no longer home. It is well that we, would-be oysters as we are, cannot rid ourselves of our feathered souls. It is well to be able to indulge in the luxury of discomfort sometimes, even if it is only to keep up our knowledge that comfort, shut up in a shell, is not a very noble thing.

VIII.

AMONG THE SUNFLOWERS.

I HAD not paid a visit to the Temple since, ten
years ago, I was a fellow guest with the curate of
my parish and his eldest daughter at a lunch party
in the chambers of my friend Jack Limpet. They
were on a visit to London in the month of May,
and I think that the young lady enjoyed Palm-
court better than Exeter Hall. Yesterday I had
to plunge into its labyrinths again, alone.

The blossom into which Clytie was turned by
her too great love for the patron of the Muses
must have been the earliest sundial. I do not
remember that there was any muse of law, though
Themis has always been on terms of cousinship, if
not of sisterhood, with the daughters of Memory.
But I do know that the courts of the Temple are

remarkable for the number of those puzzling horo-
logical instruments into which Clytie's golden dials
afterwards developed themselves.   It is true that
on my visit yesterday I found a clock on the turret
of the new hall : but I was to some extent con-
soled by hearing the dinner hour proclaimed as of
old by the primitive and unmelodious cow's horn.
Tempora mutantur is the commonest of posies for
the face of a sundial.   The glory of the Temple is
more than half departed with its walk by the edge
of the Thames.   The older it grows, the more it
becomes new.   It is no more the Temple of Goldsmith, of Charles Lamb, or even of Thackeray,
than of the knights of the Horse or Lamb.   If
Ruth Westwood, née Pinch, is still living, and
were she to make a sentimental journey to the old
fountain, she would make it in vain.   Well, *sic
aqua est*—another motto for a sunflower in brick
or stone.   It is something, in these days of progress, that the Temple has not been swept away
in steam.   But, after all, if "times are changed,"
"*non nos mutamur in illis*"—we have *not* changed
with them.

The Temple is the basilica of melancholy.

8

Burton might have made an exhaustive analysis
of his favourite science without stirring out into
Fleet-street.   It laughs loudly, it eats and drinks
hard, it is magnificently idle, it works desperately.
It is not the abode of all the virtues, but the great
virtue it has of living its life intensely.   Such is
the fascination of this intense life that men who
have once lived in it cling to it with a kind of
feline fondness.   They attach themselves to its
bricks as if they were barnacles on a ship's side.
They cling, though it is the grave of as many
hopes and good intentions as it contains paving-
stones.   Ah, here at last is No. 23, Palm-court,
with as many names upon its door-posts as if a
colony of rabbits had learned that three inches of
white paint may be let for ten pounds a year.
Here once more is Jack Limpet's name.   That
tall, broad-shouldered fellow who entertained Alice
and her father and myself so hospitably, and was
so profuse with his high spirits and champagne.
He might have made his fortune in Queensland
thrice over in these last ten years.   He might
have become the husband of Alice and of a con-
stituency too, had he pleased.   But he dropped

into this bed of sunflowers. He had dipped into "the lives of the Chancellor's" and turned towards the woolsack as Clytis to the sun. He sits from ten till four in a cupboard attached to our *Salle des pas perdus*, bandying chaff that passes for wit, and listening to flights of forensic dulness. He is not paid a shilling for his trouble: but he thinks his presence indispensable. He reads the Law Reports regularly, and does nothing else all day. Is not law notoriously a jealous mistress, and do not attorneys always avoid the man who has more than one iron in the fire? If only that one iron were a magnet! And this life he has been leading ever since the lunch party ten years ago—ten years spent in trying to keep the wolf on the other side of the oak. This is the life that he calls following a profession. It is not a profession—it is a habit, and a miserable habit besides. And this, and worse than this, is the way in which hundreds and thousands of men have spent and are spending the best part of their lives—in which scores and hundreds spend the whole. Even if fortune or a lucid interval saves him from the worst bitterness of the end of a life-long failure—even if this hope-

less strife against hope by chance succeeds, it will
be too late. Jack may possibly become Sir John ;
but I do not think that on her tomb will Alice be
styled "Dame." His red bag will have come to
contain his heart, and the means of happiness will
have become the end. "*Pereunt et imputantur*,"
says the dial opposite his chambers—"The hours
perish and it is to us they are imputed." I think
that I have at last discovered the saddest sight of
all. It is "call night" in the Temple — in a
garden of hopes which strain after a sunshine
never gained until sunset, and seldom gained even
then.

IX

AMONG THE ROOKS.

THERE are rooks and rooks. There is the rook that preys upon the farmer's wheat, and the rook that preys upon the human pigeon. The latter is common enough among us : he is the true *Corvus Londinensis.* But what of these half-dozen black rooks in their nests among still blacker London trees? There is no wheat, late or early, whereon to fatten within many a mile of Holborn Bars. Unless they can feed on red tape, or on the crumbs of not too early breakfasts, they ought to starve.

Not everybody is aware that there are real rooks in London still. That fact in natural history is known by experience only to the tenants of chambers on the west side of Gray's-inn-square.

Some zealous sportsmen have actually tried upon
these rusty wings their skill in the use of the air
gun, to the wonder of a *savant* below, who rushed
to his desk and wrote a treatise, afterwards
read before some learned society, " On the Fall-
ing Sickness among Rooks and Crows." Their
existence should also be patent to the loungers in
Gray's Inn Gardens, only—like the snakes in
Ireland—there are none. The rooks are a race
of Alexander Selkirks—monarchs of all they
survey.

It is a weekday, and within a stone's throw of
the busiest part of London : and it is more solitary
than Wimpole Street on a wet Sunday. And yet
it is not solitary. Have you quite forgotten that
you were once a schoolboy, and knew how to
climb a tree ? If not, then clamber up this elm,
and see what is to be seen. You will not see a
living form. But you will see many a living soul.
That is Lord Chancellor Bacon, who planted the
very tree on which you are perched, seeking in
vain for the summer-house from which he used to
enjoy the view of Hampstead and Highgate : you
cannot catch a glimpse of them, even from your

tree. Lord Burleigh meets him : he, too, lodged in the Inn — and shakes his head more profoundly over what we call our "statesmanship" than he did when he heard the first tidings of the Spanish invasion. Things are altered, he thinks, since the days of Good Queen Bess, whose "glorious and immortal memory" is still solemnly drunk by barristers and students in Spanish wine over the tables of Spanish oak—relics of the wreck of the Armada—that she gave to the hall. But what strange figures are these? As we rooks live, 'tis Puck and Peas Blossom, Oberon and Titania, Theseus and Hippolyta, Helena and Hermia, and Bottom the Weaver. At their head is he of the high, bald forehead and pointed beard—their poet and ours. They sweep over the rank sward, leaving fairy rings—past the chapel, on to the hall. Yes, you may forget the statecraft of Burleigh, the wisdom of Bacon, but you cannot forget that this inn of court, where the rooks still caw like lawyers at Nisi Prius, beheld the first performance of the *Midsummer Night's Dream.* We are among deed-boxes, dust, and debt, but we are in Fairyland.

Once a romance had its rise in those ground-floor chambers now occupied by some artist, journalist, or attorney—by any one but a barrister, for the bar fights shy of this corner of legal Bohemia. One fine summer's evening a party of students sat at the window over their wine. A pretty nursemaid, airing her charge in the garden —more frequented then—was not averse to a little conversation. By and by she was prevailed upon, by way of jest, to hand the infant into the room. She was seen to pass round the corner on her way to reclaim it at the door, but was never seen again. The young men found themselves in the joint possession of a fine little girl of some nine months old. Making her a sort of "Figlia del Reggimento," they bound themselves to be thenceforth the protectors and godfathers of her whom they christened Emma Gray. Is not that an idea for the beginning of a novel? She might be a duke's daughter, and in the last chapter marry the handsomest of her godfathers. What an opening for a plot there would be—one young girl forming a link to bind together a dozen men with hostile interests and conflicting characters. It is the fitter subject for romance, for of her after life history tells us nothing. Only one may be sure that the cawing of rooks sounded all her life long like a melody running through the discordant London roar. That unmusical music is full of vague, suggestive, dreamy harmonies when it sounds among green fields and sunny hills. How much more suggestive is it of contrasts and of memories when it grates upon the ear in that oasis of silence in a desert of sound which we call Gray's Inn.

X.

AMONG THE CAGES.

THE whole natural history of London may be read in its ornithology. To begin with, every grade of society is represented by its feathered population, caged or uncaged, from the sparrow that gains its living from the streets, in the face of the watchful constabulary of the tiles, to the almost royal parrot, who has a whole household at command, and is treated as though he were a prince bewitched by an evil genii, or like some popular statesman whose random words, though repeated for the thousandth time, are gathered up as though they were pearls. Every race is typified by our fellow biped, from our own native finches who, in captivity, become the *bourgeoisie* of British birds, to the gorgeous strangers who cross the sea

from Malacca or Peru. Again, they have their
proper quarters of the town. Just as one goes to
Whitechapel to look for a German, to Soho for a
Frenchman, and to Hatton Garden for an Italian,
so do the thrushes congregate in St. Giles's, the
rooks by Holborn, and the canaries in Spitalfields·
Of individual character there is no need to speak.
They are as various as among women—who are
not so very various, after all. Probably thief does
not differ from thief much more than sparrow from
sparrow, or more than lawyer from lawyer, crow
from crow. All are leaves of the world's tree,
which differ in microscopic detail, but grow and
live and wither in the same seasons and by the
same law.

People may say what they will about the
weight of bricks and paving-stones that imprisons
the human heart of London, which must be
beating somewhere, even though it be so hard to
read, so hard even to find. Why, those who have
ears to hear through the rattle and whirl of the
streets cannot listen for a single instant without
catching the music of its pulsations. Buried deep
it may be, but its very depth renders it only the

more deeply vocal. The soprano song of the woods, when caged, only turns to the contralto voice which contains the true pathos of song—the voice which comes from the imprisoned heart, not only from the throat which sings just because it was made to sing. It is but that we ourselves have no time, perhaps no power, to sing. We have to free our hearts by deputy, as the Japanese are said to pray. Only we may be sure that the song is in our case none the less sweet and musical for being caged, or for our deputies being the innocent singers that they are. It is fitting that angels should be represented with feathered wings.

I have said that canaries are a feature of Spitalfields. It must surely be by force of contrast if it is by force of any reason at all. Indeed, the love for this special kind of birds amounts to a passion, for which some people strive to account by tracing a subtle connection between the descendants of the exiles of the Azores and the descendants of the exiles of Auvergne, who muster in that quarter strongly still. There may be something in such sympathy:

but, be this as it may, yonder old fellow who is slouching round the corner has not a drop of French blood, however remote, in his veins. He is neither a weaver nor the great grandson of a weaver. He is prematurely broken down by misery and gin. He spends half his days in public-houses, the other half in prison-cells. He never knew a letter of the alphabet, but he has been a good workman in his time, though never a sober one. But in those days, though he never had a wife or child, or knew father or mother, he had a home ; and solely to keep that home together he toiled hard during his sober hours and kept himself from crime. He literally lived for a fellow-creature, and that fellow-creature was a bird. He had to keep a roof over its head, and to save sufficient halfpence from the barman's till to keep it in health and luxury. Does that seem incredible ? To him it seemed the most natural thing in the world—and not only natural to him, but to many of his neighbours too, who were by no means a colony of fine ladies or tender-hearted girls. He had bought the bird for a bargain from a canary-fancier, who was a broken-down pugilist :

but the golden notes of its song somehow pre-
vented his trying to sell it again. It seemed to
penetrate among the unconscious possibilities of
what might have been. He even fought battles in
its honour in the spirit, if not according to the
forms, of ancient chivalry. It was thus that he
fell at last into the clutches from which he had so
long tried to keep himself free. A street fight led
to a week or two in gaol : and when he came out
again the bird had long been starved. His one
song was silent ; and he himself became the caged
gaol-bird that you see.

Such are London birds, that learn in captivity
to sing for us our hidden songs, and to build their
nests in hearts instead of leaves.

XI.

AMONG THE WATER-LILIES.

A STUDENT of London's natural history who, in order that his work may be complete, feels himself bound to devote a chapter to its water-lilies, is in the position of the celebrated naturalist who wrote that celebrated chapter upon the snakes of Ireland. Thanks to St. Patrick, the chapter was unique for brevity. It is needless to add that, by one gazing from any of the river bridges, no water or any other lilies are to be seen, unless there be a Citizen or Iron boat bearing that name. But there was once, in days before London was dreamed of, a certain goddess, well known at Memphis and Alexandria. Her name was Isis. She wore a garland of lotus-flowers, which are nothing less or more than the water-lilies of the Nile, which was nothing more or less than Cleopatra's Thames.

She was veiled from head to foot, and that veil never was lifted by any profane hand. Whether she was a barbaric Venus, or whether she was like the skeleton at her country's feasts, no one but her priests could say. London is the Isis among cities ; and for that reason, whether the Thames bear water-lilies or no, we are among the water-lilies, whatever they may be, that form her garland. She, like her mythological prototype, is veiled : nor are there many eyes able to pierce below her rags and her fine clothes. But once in the twenty-four hours she throws off her disguise : and the hour she chooses for her self-revelation is when she thinks there are no eyes to see. Indeed, it is more true to say, not that she throws off her veil, but that her veil falls from her when she is sleeping among the lotus flowers of dreams—the only kind of lotus-eating our days, wherein it is never after-noon, permit us to know.

I have spoken of one hour in the twenty-four. Rather should I have spoken of one half-hour in the eight-and-forty. It is for not more than thirty short and sudden minutes that London is freed from rags and robes. It is not more than thirty minutes that stand between the last man who goes to bed at night and the first who rises in the morning, between the latest bird and the earliest worm. Few people know there is such a time at all. The theatres are over long ago. The earliest market waggons have not yet passed out of sight of the gardens that form the fringe of the veil. Even the ball at which you have been playing at enjoying yourself has received its apotheosis in the last after-supper galop. Say that it has been in the south-western region, built upon marshes in whose midst the abbey was once a solitary island as well as a sanctuary, where houses stand for long-forgotten water-lilies. You are the worse for want of oxygen—for nothing more. For once let us be superior to the temptation of waking from his sleep on the box the driver of what Mr. Disraeli was once pleased to call a gondola. The term is more appropriate than ever. The canals of Venice, the dead city, are not so silent, so solitary, so dead, as these London streets, fresh from the builder's hand. You, in dress-clothes, have suddenly left behind an atmosphere in which you vainly thought yourself admirably clothed and in your sane mind, to pass the threshold of a world in which you feel like Oberon in a drawing-room. The tables are turned.

The outlines of stucco are as clear and sharp in the grey light as if new chiselled in marble by the hands of fairy tomb-sculptors. No kitchen fires as yet are lighted to blear them into strokes traced on damp blotting paper. You feel, for once, what it is to be alone in an unknown and unrecognised world. You are a stranger in the streets which are as familiar to you by day as your own staircase, or the country lanes in which you used to ramble once upon a time. There is something in the atmosphere that is altogether new and strange. If it is your fate to pass through St. James's Park, you can hardly find your way. The pond blossoms with imaginary water-lilies : even Buckingham Palace looks imposing. In the flesh you are at home, in the spirit you are a stranger. In the streets you notice names, numbers, combinations of doors and windows, that you never noticed before. All things have suffered a change in the uniform grey air. The pavement, trodden by so many heels, devotes itself to multiply your

own footsteps a hundred times, until you feel pain-
fully as though there were nothing in all London
but you, and as though yourself were changed into
sound. Even the two policemen, always to be
found conversing at stray corners where no tres-
passer but a cat ever comes, and whose tramp, as
they part, sounds like a muffled repetition of your
own rapid tread, take upon themselves an appearance of phantoms. You would laugh at the thought
were laughter possible in an air in which you fear
to wake even the echoes. If you are of a nervous
temper, no dream that the early lateness of the
hour will permit you to steal from morning will
feel so like a dream. You are fatigued in body,
and morbidly unfatigued in mind. You are awake
and asleep. Yours are the only waking eyes, the
only moving limbs, in a world of life. Your brain
becomes morbidly conscious of the slightest sight,
of the slightest sound : while nothing is to be seen
but a vista of straight lines—nothing to be heard
but your own footfall. London has grown mysti-
cally beautiful—her veil has fallen. But—what is
that primrose tint, creeping up through, not along,
the sky ? You have a hundred times witnessed the

saffron glow of a fire. But this—it is sunrise!
Yes, the sun rises, even in London. Hurry home-
ward, for the charm is about to fade. That is a
spiral of smoke curling into the air, to blot out the
first beam of the dawn. Isis is waking. She is
donning her weeds, that stand to her for lotus-
flowers.

There is nowhere but in the one half-hour of
London's sleep that a man can feel himself to be
alone. Between the hedges he hears the song of
birds, even though it be reduced to the hoot of the
owl. Nature never sleeps—not even among her
alps and seas. London sleeps for one short half-
hour, but in that short half-hour she turns all
things into a dream. Life is her veil. When life
sleeps she wakes, and becomes, not the metropolis
of the world, but of the land in which Oberon is
king—king even over those who dance by night
and hide themselves from her in the counting-
house by day. For thirty minutes London lives
without life, and teaches those who know her not
how little she is known.

XII.

AMONG THE SPARROWS.

WHY has no•painter, without funds to carry him far afield in search of novelty, made a reputation by turning into a picture a subject that he may find without going more than a mile or two—perhaps not a hundred yards—from home? The broad walk of St. James's Park, as we late risers know it, is not picturesque. In its recognized hours it is a parade ground for cabs, perambulators, guardsmen, and cockneys. It has its uses : but the useful is not of necessity the beautiful. But there is one hour of the four-and-twenty of a spring day when it is absolutely beautiful. And that hour is when the sun first rises, and the sparrows hold clamorous parliament before they set out upon the chase of the early worm.

Before the kitchen fires are lighted, not even the boasted atmosphere of Paris or Florence is clearer, brighter, or fresher than ours. So pure are then the outlines of our dullest streets, that our palaces look almost imposing, and our monuments almost like the works of art that they pretend to be. The new-leaved trees, glittering with sunlit dew, show no symptoms of town save in their blackened stems. Through them, its grey walls tinged with rose, break glimpses of the abbey. Even the barbarous clock-tower adds to the effect when its gilding reflects the morning's gold. It is a scene in which the very sparrows, cockneys as they are, seem to sing a country song.

But it is this terrible London of ours, after all. The sparrows' chatter dispels a charm. They are the destroyers of dreams. You, who are drinking in the freshness of morning, are either outwatching the night or anticipating the day. But in either case you are the owner of a bed from which you have just risen, or in which you are about to lie .

down. Not so those who come back from dream-
land into life at the summons of these birds. As
you pass, the benches wake up slowly, and yield
again to the streets what they had gathered from
the streets the night before. Who are all these—
whence do they come, and whither do they go?
There is no sadder sight than to see how the
beauty of nature may for an hour triumph over
an atmosphere of bricks and mortar, but cannot,
even for an hour, hold its own against the misery
for which there are no bricks and mortar to build
a home. That is no common pauper who is just
stretching out his ill-rested limbs. Do you remember Rockett of St. Margaret's — the most
brilliant scholar and best dressed man of his year?
Some labelled him bishop : some prime minister.
Well—that starved, blotched, ragged fellow is he.
Is it drink that has brought him to this? Is it
dice? Is it overwhelming fate? Pass by on the
other side, unless you are so greedy of gratitude as
to wish to be abjectly thanked for the price of a
roll at the nearest coffee-stall. On the same bench
still sleeps, despite the sparrows, a man who was
once a millionaire, a director of railways and

mines. But he has gone under the waves of life—
to-morrow, look for him under the waters of the
Thames. On the next is one who came to town
six weeks ago to turn his brains and 20*l.* into
gold. He was poet, painter, musician, actor—
what matters what he was in his own conceit when
he has spent his last penny, and is neither energetic
enough nor rogue enough to turn another? He
wakes from dreams of fame to find himself between
a forger flying from the police, and a drunken
cripple. None of these, except the last, but is
prouder than the refuse of the casual ward. But
not far off is one who is in a worse plight still :
for it is a woman. May she not wake too soon to
her daily life, unless the dreams of home with
which the sparrows' twitter fills her sleeping ears
are too terrible to bear.

No : it is no wonder, after all, that no painter
has made this his out-door studio. He would be
more of a cynic than it is in the nature of a true
artist to be, were he able to fix his mind upon the
glittering leaves, the fresh air, and the distant
abbey that is our historic sanctuary. He could
not have the heart to paint for wealth or fame in

an atmosphere of failure, disappointment, and despair. Perhaps not one of those around him may chance to be the hero or heroine of a true tragedy : but there is none, drunk or sober, who is in sympathy with the breeze of daybreak, or the merry chatter of the birds.

THE END.